Alexandru Dimca

Topics on Real and Complex Singularities

Advanced Lectures in Mathematics

Edited by Gerd Fischer

Jochen Werner
Optimization. Theory and Applications

Manfred Denker
Asymptotic Distribution Theory
in Nonparametric Statistics

Klaus Lamotke
Regular Solids and
Isolated Singularities

Francesco Guaraldo, Patrizia Macrì,
Alessandro Tancredi
Topics on Real Analytic Spaces

Ernst Kunz
Kähler Differentials

Johann Baumeister
Stable Solution of Inverse Problems

Alexandru Dimca
Topics on Real and Complex
Singularities

Alexandru Dimca

Topics on Real and Complex Singularities

An Introduction

Springer Fachmedien Wiesbaden GmbH

CIP-Kurztitelaufnahme der Deutschen Bibliothek

Dimca, Alexandru:
Topics on real and complex singularities: an introd. /
Alexandru Dimca. – Braunschweig; Wiesbaden:
Vieweg, 1987
(Advanced lectures in mathematics)

ISBN 978-3-528-08999-3 ISBN 978-3-663-13903-4 (eBook)
DOI 10.1007/978-3-663-13903-4

AMS Subject Classification: 14 B 05, 14 C 20, 14 E 15, 14 H 20, 14 J 17, 14 L 30, 14 M 10, 32 A 05, 32 B 05, 32 B 10, 32 B 30, 32 C 40, 32 C 45, 58 A 20, 58 C 25, 58 C 27.

Originally published by Friedr. Vieweg & Sohn Verlagsgesellschaft mbH, Braunschweig 1987.

Produced by Lengericher Handelsdruckerei, Lengerich

To My Parents

TABLE OF CONTENTS

INTRODUCTION

The body of mathematics developed in the last forty years or so which can be put under the heading *Singularity Theory* is quite large. And the excellent introductions to this vast subject which are already available (for instance [AGV], [BG], [Gi], [GG], [Lm], [Mr], [Ws] or the more advanced [Ln]) cover necessarily only a part of even the most basic topics.

The aim of the present book is to introduce the reader to a few important topics from *local* Singularity Theory. Some of these topics have already been treated in other introductory books (e.g. right and contact finite determinacy of function germs) while others have been considered only in papers (e.g. Mather's Lemma, classification of simple 0-dimensional complete intersection singularities, singularities of hyperplane sections and of dual mappings of projective hypersurfaces). Even in the first case, we feel that our treatment is different from the introductions mentioned above - the general reason being that we give special attention to the *complex analytic* situation and to the connections with *Algebraic Geometry*.

We offer now a detailed description of the contents, pointing out special aspects and new material (i.e. previously unpublished, though for the most part surely known to the experts!).

Chapter 1 is a short introduction for the beginner. We recall here two basic results (the Submersion Theorem and Morse Lemma) and make a few comments on what is meant by the local behaviour of a function or of a plane algebraic curve.

In Chapter 2 we introduce the local algebras E_n of germs of smooth or complex analytic functions and study some of their basic algebraic properties. We state (but do not prove!) Nakayama's Lemma (2.8) and the analytic Hilbert Nullstellensatz (2.14) which are used several times in what follows.

Next we list some of the *main differences* between the real smooth and the complex analytic cases in Proposition (2.15) and Remark (2.16).

At the end of this chapter we introduce the jet spaces which are among the main finite-dimensional objects of the theory.

Chapter 3 starts with a general discussion on equivalence relations induced by group actions. We define then some infinite dimensional groups (of germs of diffeomorphisms and of invertible matrices over E_n) and, using them, we introduce the basic equivalence relations on map germs: right equivalence $\mathcal{R}$, left equivalence $\mathcal{L}$, right-left equivalence $\mathcal{A}$ and contact equivalence $\mathcal{K}$.

Next we bring in *germs of analytic spaces* (alias local analytic $\mathbb{C}$-algebras) and relate several natural notions of isomorphism for such germs to the contact equivalence in Proposition (3.16). (In this book we let $\mathbb{N}$, $\mathbb{Z}$, $\mathbb{R}$ and $\mathbb{C}$ denote the natural, integer, real and complex numbers respectively).

Finally in this chapter we show that the infinite-dimensional group actions (used to define the equivalence relations listed above) are compatible with the process of passing to jet spaces. They yield in this way some algebraic group actions which turn out to be good finite-dimensional approximations to the original actions.

We begin Chapter 4 by recalling some basic facts on semialgebraic (alias constructible) sets. To give the reader a correct idea about what a singularity is, we include here a simple remark of Milnor [M1], to the effect that the germ of an analytic set in $\mathbb{C}^n$ is a real differentiable submanifold in $\mathbb{C}^n$ only when this germ is smooth (i.e. nonsingular) in the complex analytic sense, see Proposition (4.4). Note the *two distinct meanings* of the word "smooth" in our book: C^∞-differentiable and nonsingular.

We use semialgebraic sets mainly to show that the orbits of an algebraic action (on a smooth variety) are smooth semialgebraic submanifolds. Next we explain how one can compute the tangent spaces to such an orbit and give several concrete examples.

We introduce the basic notion of a transversal (alias slice) and explain its role in understanding the local structure of a smooth action. Then we compute the tangent spaces to the $\mathcal{R}$, $\mathcal{L}$, $\mathcal{A}$ and $\mathcal{K}$ orbits in a jet space.

We end this chapter with the statement and proof of *Mather's Lemma* (4.26). Since in recent years this result has been used more and more (and since we need it several times in our book!) we consider its inclusion in a book of introductory nature like ours completely justified.

In Chapter 5 we present some simple, classical and (we hope) instructive examples of the classification of the orbits for some algebraic actions. These actions are very close to the actions on jet spaces introduced before. On the other hand, they offer us the opportunity to show concretely the importance of computing tangent spaces and slices to the orbits and to introduce the basic concept of *specialization* (*adjacency*) in the simplest interesting situations.

We discuss in turn the classification of linear maps and of quadratic and cubic forms. This latter case has a different level of difficulty and brings in substantial connections with Algebraic Geometry. Perhaps a new fact here is our description of *all the specializations among projective plane cubic curves* (5.17).

Next we discuss (quickly) the pencils of quadratic forms and (in detail) the *pencils of binary cubic forms* (5.22). The latter seems again to be new and is quite relevant to the classification of map germs $(\mathbb{C}^2,0) \to (\mathbb{C}^2,0)$ as shown in [DG2], [DG3], where this classification of pencils of binary cubic forms is used without proof. In addition, this is one of the few cases where a complete listing of the orbits is possible and has a nice geometric interpretation, see Figure (5.24).

With Chapter 6 we reach the heart of the topics discussed in this book, namely the finite determinacy of germs. From this point on we concentrate only on the R and K equivalence relations.

There are three main reasons for this limitation:

(i) the treatment in these cases is similar and much simpler that in the L or A equivalence case.

(ii) historically, the case of R equivalence was considered first in doing explicit classifications (e.g. Arnold's classification of R simple function singularities).

(iii) the case of K equivalence is the most important for

connections with Local Analytic Geometry and Algebraic Geometry (see (3.16), (6.39), (6.42) and the whole of Chapter 11).

We treat first the finite $\mathcal{R}$ determinacy of function germs $(K^n,0) \to (K,0)$ with $K = \mathbb{R}, \mathbb{C}$. Here we introduce a notion (*strong determinacy* (6.8)) which is a more precise version of the k-open orbit in [Mr], p.27 and which will play an important role in what follows (e.g. in the proof of Theorem (11.41)).

Next we give *two proofs* of the basic Theorem (6.11) on finite $\mathcal{R}$ determinacy. The first proof is limited in its range of application to the complex analytic case and is technically more difficult, involving *Artin's Approximation Theorem* (6.14) and Mather's Lemma. However, this proof has the advantage of being a very clear step-by-step construction, quite general in spirit and likely to be useful in a lot of similar situations (e.g. we use it in the proof of the corresponding result for finite $\mathcal{K}$-determinacy (6.27)).

The second proof (by showing the triviality of a 1-parameter deformation) is the standard one for this result.

Then we study the finitely $\mathcal{K}$-determined map germs $(\mathbb{C}^n,0) \to (\mathbb{C}^p,0)$ and, for $n \geq p$, we relate them to the *isolated singularities of complete intersections* in Proposition (6.39).

Next we say what a generic property (in the sense of Tougeron) is and show that the property of being finitely $\mathcal{K}$-determined is generic (6.46). And we prove that a function germ $f:(K^n,0) \to (K,0)$ is finitely $\mathcal{R}$-determined if and only if it is finitely $\mathcal{K}$-determined. The complex case is easy to prove, but to derive the result in the real case we use an idea of Wall [W2] and bring in a deep *theorem of Briançon-Skoda* (6.47), (6.48).

Then we introduce the Milnor number $\mu(f)$ and the Tjurina number $\tau(f)$ for a finitely determined function germ f and for the corresponding isolated hypersurface singularity $(f^{-1}(0),0)$. Next we prove an interesting and apparently new result, namely that the *$\mathcal{R}$-sufficient orbits contained in a $\mathcal{K}$-sufficient orbit* $\mathcal{K}f$ *form a foliation (in a sufficiently large jet space), the codimension of the leaves being exactly* $\mu(f)-\tau(f)$, see Corollary (6.55).

In Chapter 7 we first prove the equivalence between two natural concepts of weighted homogeneous singularity: weighted homogeneous defining equations and invariance with respect to a corresponding $\mathbb{C}^*$-action. Then we treat the finitely determined weighted homogeneous map germs and establish the well-known formula for the Milnor number in terms of the weights and the degree (7.27) using the Poincaré series associated to graded algebras.

Then we briefly treat the semiweighted homogeneous singularities and present a *reduction to normal form for such a map germ* (7.41), which is usually stated only for function germs as in (7.38).

Next we discuss (with incomplete proof) a basic result of K. Saito giving conditions under which a function germ is equivalent to a weighted homogeneous polynomial, see Theorem (7.42). Moreover the weights of such a polynomial can be normalized in a standard way (7.43) and this leads us to consider, for the first time in our book, the normal forms A_k, D_k, E_6, E_7 and E_8 defined by a numerical inequality, see Corollary (7.45).

We end this chapter by describing a uniqueness property of the weighted homogeneity type of isolated complete intersection singularities (7.47), due essentially to Wall [W1].

Chapter 8 starts with a discussion of simple singularities and specializations. Here we formalize a notion (*sufficient neighbourhood* (8,5)) which is quite useful and which is implicit in most of the literature devoted to this subject.

Then we present an efficient procedure for doing K-classifications of map germs (the use of *complete transversals*) which was introduced and used systematically by Gibson and the author [DG3], [DG4].

We use this method of complete transversals to give a proof of the Splitting Lemma (8.12) and to derive the classification of the K-simple function singularities, leading to the famous A-D-E singularities, see Theorem (8.26).

Then we obtain the classification of the R-simple function singularities in two distinct ways:

(i) using the classification of the K-simple function singularities;

(ii) using a basic (seemingly new) remark that *one can prove a priori that an R-simple function singularity is weight-*

ed homogeneous and then using the characterization of the normal forms A_k, D_k, E_6, E_7 and E_8 by a numerical inequality as mentioned above. This last proof is related to Arnold's concept of *inner modality*, [AGV], p. 218.

Chapter 9 deals with the classification of K-simple map germs $(\mathbb{C}^n,0) \to (\mathbb{C}^n,0)$. The corresponding normal forms were obtained by Giusti [Gt], but we use a more convenient method for deriving them (again the complete transversals!) and, more importantly, we give also *the list of all the specializations among these singularities*. These results are joint work of Gibson and the author and are contained in the unpublished preprint [DG1] (the statements without proofs can also be found in [DG2]).

Apart from this, Chapter 9 contains some useful general material on the Boardman symbol and the Hilbert-Samuel function of a map germ (or of an ideal $I \subset E_n$).

In Chapter 10 we study in some detail the isolated hypersurface singularities of dimension one and two.

First we give a *new proof* of the Weierstrass Preparation Theorem (10.4), valid only for finitely determined function germs and using only the complete transversal method.

Next we start the study of plane curve singularities with Hensel's Lemma (10.8) and discussions on local intersection numbers and Puiseux parametrizations.

Then we associate a *new numerical invariant* $k(X,0)$ to a plane curve singularity $(X,0)$, related to the Boardman symbol of a defining equation for $(X,0)$. And we study the behaviour under one blowing-up of this invariant $k(X,0)$, of the multiplicity mult $(X,0)$ and of the Milnor number $\mu(X,0)$.

These results lead naturally to a proof of the embedded resolution of the plane curve singularities, see Theorem (10.30).

Next we give a characterization of the simple plane curve singularities A-D-E in terms of the multiplicities of the singularities of the reduced total transform (after one blowing-up) as in [BPV].

In the second part of this chapter we turn our attention to the simple surface singularities and discuss some of their basic (and characteristic) properties. In particular we relate them via their minimal resolutions to the classical Dynkin diagrams A_k, D_k, E_6, E_7 and E_8 associated to the simple Lie groups (Lie algebras) whose root systems have only roots of equal length.

We have given in great detail an *explicit construction of the minimal resolutions for the* A_k*-singularities using only blowing-up of points*, a basic result for which we have not found a proper reference at this elementary level.

One use for the Dynkin diagrams (apart from explaining the labels A-D-E attached to the simple hypersurface singularities) is described in Remark (10.55), where we relate the specializations of simple singularities to the inclusions of their associated Dynkin diagrams.

Chapter 11 gives some applications of the notions and results of Singularity Theory presented in the book to a natural problem in Algebraic Geometry: the study of *the singularities of the dual mapping* of a complex smooth projective hypersurface V. Equivalently, this is the study of the singularities of the *hyperplane sections* of the hypersurface V or, in other words, the study of the *contact* between the hypersurface V and its tangent hyperplanes at various points. This "contact" is formally represented by a contact (tangency) class of a function with an isolated singularity. It is shown that this class determines the contact class of the dual mapping germ at the corresponding point, see Proposition (11.30).

Next we characterize the simplest contact tangency classes, namely those of type A_1 in (11.33) and A_2 in (11.34) - this latter case being new as far as we know.

The main result of the chapter is Theorem (11.41) which describes the *generic contact tangency classes* when one fixes the dimension and the degree of the hypersurface V. This result is a new version (more appropriate for the algebraic geometers in our opinion!) of a nice result of Bruce [Bc].

Although the material in this chapter is essentially elementary and perhaps well-known to the specialists, we have presented it here for two reasons:

(i) it is difficult, if not impossible, to find a proper

reference for most of it;

(ii) we hope that these simple applications will attract some algebraic geometers (resp. workers in singularities) to the wonderland of Singularity Theory (resp. Algebraic Geometry), to the benefit of both fields.

To introduce the reader to *current research literature*, we have included a short Appendix, in which we mention two topics of recent interest, closely related to the topics discussed in the book.

The reader is expected to have a basic knowledge about smooth manifolds (and a few times about Lie groups) as well as a minimal training in (Linear and Commutative) Algebra and in (Algebraic and Local Analytic) Geometry. In spite of this, we have included the definitions and the statements of the most important notions and results which we have used.

We assume from the reader no previous contact with Singularity Theory, but he will find it perhaps profitable to read parallel sections in [Gi], [Mr] or [AGV]. On the other hand, the reader who is already familiar with some aspects of Singularity Theory may find here new points of view and complementary results, because of our concern for the complex analytic setting.

Many *exercises* are scattered throughout the book and they are a good (and usually simple!) test for the reader's understanding of the subject. The book also contains a few problems which, as far as the author knows, are still *open* and of interest.

My outlook on singularities has greatly benefited from discussions or correspondence with many mathematicians to whom I am deeply grateful. In my early (and maybe most difficult) stages of development I have received invaluable help and advice from C.G. Gibson. Later on, my knowledge was deepened and my enthusiasm for the subject was enhanced through my contacts with: V.I. Arnold, E. Brieskorn, J. Damon, A. du Plessis, W. Ebeling, A.B. Givental, G.-M. Greuel, H. Hamm. Lê Dũng-Tráng, K. Saito, D. Siersma, J.H.M. Steenbrink, A.N. Varchenko and

C.T.C. Wall.

Special thanks are due to S. Dimiev who encouraged me to write a book like the present one and to W. Ebeling who suggested that I publish it with Vieweg Verlag. And to my friends D. Popescu and G. Teodosiu who have read parts of the manuscript and helped me to avoid some mistakes.

I am also very grateful to Mrs. Ulrike Schmickler-Hirzebruch from Vieweg for her kind attention and efficient help in preparing the manuscript. And to Miss Camelia Minculescu who has succeeded in producing a reasonably good typescript under rather difficult circumstances.

Finally it is a pleasure to acknowledge my great debts to my wife Gabriela, who has supported me in various ways, a major one being to prevent our two noisy little sons from playing with me all day long.

Bucharest, April 1987 Alexandru Dimca

CHAPTER 1

TWO CLASSICAL EXAMPLES: SUBMERSION THEOREM AND MORSE LEMMA

For a natural understanding of the notions to be introduced in the next chapters, we recall here two elementary results which are basic in many parts of Geometry and Topology.

(1.1) SUBMERSION THEOREM

Let U *be an open neighbourhood of the origin* $0 \in \mathbb{R}^n$ *and* $f: U \to \mathbb{R}^p$ *a smooth map such that*

(i) $f(0)=0$

(ii) f *is a submersion at* 0 (*i.e.* rank $df(0)=p \leq n$). *Then there is an open neighbourhood* $U_1 \subset U$ *of the origin* $0 \in \mathbb{R}^n$ *and a diffeomorphism* $g: U_1 \to \mathbb{R}^n$ *such that*

(a) $g(0)=0$

(b) $f \circ g^{-1}(x_1,\ldots,x_n)=(x_1,\ldots,x_p)$.

Here df(0) denotes the differential of the map f at the point 0, regarded as a linear mapping between the corresponding tangent spaces $\mathbb{R}^n \to \mathbb{R}^p$ and given by the $p \times n$ matrix of partial derivatives $\frac{\partial f_i}{\partial x_j}(0)$, with $i=1,\ldots,p$; $j=1,\ldots,n$.

Note that we can reformulate the conclusion (b) of this Theorem in a more *classical language* as follows: There exists a system of coordinates $(\bar{x}_1,\ldots,\bar{x}_n)$ around the origin $0 \in \mathbb{R}^n$ such that the original map $y=f(x)$ can be written in this new coordinate system in the form

$$y_1=\bar{x}_1,\ldots,y_p=\bar{x}_p \ .$$

On the other hand, the *geometrical meaning* is that a submersion (at a point x_o) behaves (around x_o) like a linear projection.

The proof of Theorem (1.1) is an easy consequence of the Inverse Function Theorem and the reader who needs more details can find them in the first Chapter of Gibson's book [Gi].

To state the second result, we need the following.

(1.2) DEFINITION

Let $U \subset \mathbb{R}^n$ be an open set, $x_o \in U$ a point and $f: U \to \mathbb{R}$ a smooth function. The point x_o is called a *nondegenerate singularity* of the function f if

(i) $df(x_o)=0$

(ii) the Hessian matrix of f at the point x_o is nondegenerate, i.e.

$$\operatorname{rank}\left(\frac{\partial^2 f}{\partial x_i \partial x_j}(x_o)\right)_{i,j=1,\dots,n} = n .$$

Sometimes, especially in Algebraic Geometry, such a point x_o is called a (nondegenerate) *quadratic singularity*. This is justified by the following fundamental result.

(1.3) MORSE LEMMA

Let U *be an open neighbourhood of the origin* $0 \in \mathbb{R}^n$ *and* $f: U \to \mathbb{R}$ *a smooth function such that*

(i) $f(0)=0$

(ii) 0 *is a nondegenerate singularity of the function* f.

Then there is an open neighbourhood $U_1 \subset U$ *of the origin* $0 \in \mathbb{R}^n$ *and a diffeomorphism* $g: U_1 \to \mathbb{R}^n$ *such that*

(a) $g(0)=0$

(b) $f \circ g^{-1}(x_1,\dots,x_n) = -x_1^2 - \dots - x_p^2 + x_{p+1}^2 + \dots + x_n^2$ *for some integer* p, $0 \le p \le n$, *called the index of the singular point* $x_o=0$ *of* f.

In other words, a smooth function f behaves around a nondegenerate singularity exactly like a nondegenerate quadratic form.

A proof of Morse Lemma will be given in the sequel (see Chapter 6), but some readers may enjoy the short direct proof of this result given at the beginning of Milnor's book [M2].

There is a completely similar statement to (1.3) for a complex analytic function $f:U \to \mathbb{C}$, where U is an open subset in $\mathbb{C}^n$ (we just don't need the minus signs and the image of g should be taken to be an open ball!).

In a special case, this was well-known in Algebraic Geometry for a very long time. Let F be a polynomial in $\mathbb{C}[x_1,x_2]$ and let $X=\{x\in\mathbb{C}^2;\ F(x)=0\}$ be the corresponding *affine plane algebraic curve*. Assume that $0\in X$. Then 0 is called a *node* of X if 0 is a nondegenerate singularity for the function $F:\mathbb{C}^2 \to \mathbb{C}$. If this is the case, it follows from the complex version of (1.3) that, locally around 0, the curve X is equivalent to the algebraic set given by $x_1^2+x_2^2=0$, i.e. the union of the two lines $x_1+ix_2=0$ and $x_1-ix_2=0$.

Note that globally X may be irreducible and hence its study cannot be reduced to the study of simpler objects. An example of this situation is given by the nodal cubic curve given by $F(x_1,x_2)=x_1^2(1+x_1)-x_2^2=0$

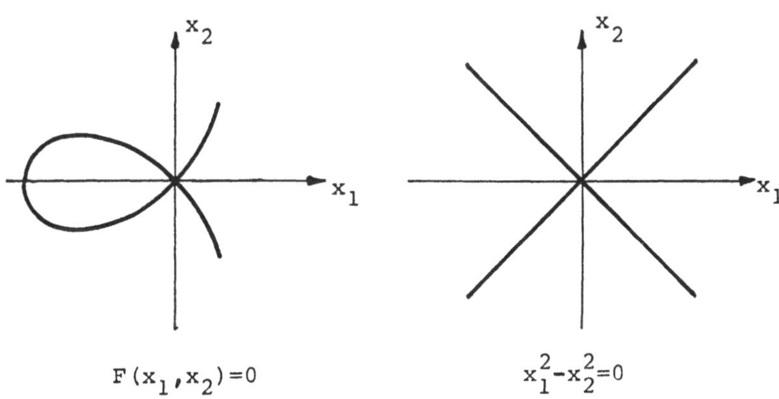

(1.4) FIGURE

(1.5) EXERCISE

Show that the above polynomial F is irreducible in $\mathbb{C}[x_1,x_2]$.

The above two results (1.1) and (1.3) have essentially the same structure: starting with a *punctual condition* (in our case in the origin $x_o=0$ of $\mathbb{R}^n$) on a smooth map f and on its derivatives of different orders, one gets information on the local behaviour of the function f (on the open set U_1). They say that, up to a coordinate change (which, by the way, is completely unknown or, at least, unimportant as is also the open set U_1), the mapping f is described by some quite simple polynomial formula, a so called *normal form*.

It is useful to think of this type of results as modern and more difficult counterparts to classical results in Linear Algebra e.g. the reduction of a quadratic form to the diagonal normal form or the reduction of a square matrix (linear endomorphism) to the Jordan normal form.

Why such results are important ?
For instance, Theorem (1.1) gives a trivial proof to the fact that the fibers $f^{-1}(y)$ of a submersion $f:X \to Y$ between two smooth manifolds are submanifolds in X.

And Morse Lemma can be regarded as the starting point of *Morse Theory* [M2], in which one studies the connection between the global topological properties of a smooth manifold X and the local structure of a generic function $f:X \to \mathbb{R}$, a so called Morse function.

In both applications, the arguments are *local* and hence we can replace the submersion (resp. the Morse function) f by the very simple corresponding normal form.

In Singularity Theory the local behaviour of a map around a point is described formally by the *germ* of the map at this point and the coordinate changes are replaced by more general *equivalence relations* among various classes of germs.

The next two Chapters will introduce the basic notions and notations necessary for handling properly these objects.

CHAPTER 2

GERMS AND JETS

§1. GERMS OF SETS AND MAPPINGS

Let X be a topological space and $x \in X$ a point. On the set $P(X)$ of all subsets in X we define an equivalence relation: $A \underset{x}{\sim} B$ if and only if $A \cap U = B \cap U$ for some neighbourhood U of x.

(2.1) DEFINITION

An equivalence class of the relation $\underset{x}{\sim}$ is called a *germ of a subset of* X *at the point* x.

Usually we shall denote the equivalence class of a subset $A \subset X$ again by the letter A, the point x being clear from the context. A more accurate notation for this germ is (A,x) and we shall use this longer form when necessary.

Let now Y be a set and consider the set of pairs $M=\{(U,f)\}$, where U is a neighbourhood of x in X and f is any function $f:U \to Y$. We introduce an equivalence relation on $M: (U_1,f_1) \underset{x}{\approx} (U_2,f_2)$ if and only if $f_1|U_o=f_2|U_o$ for some neighbourhood U_o of x with $U_o \subset U_1 \cap U_2$.

(2.2) DEFINITION

An equivalence class of the relation $\underset{x}{\approx}$ is called a *germ of a map from* X *to* Y *at the point* x. Usually we shall denote the equivalence class of (U,f) simply by f. A more accurate notation is f_x or (f,x) and these are used when necessary.

Note that we forget completely the neighbourhood U on which (a representative of) our germ f is defined. On the other hand, the *value* $f(x)$ is well-defined, since it is the same for all the representatives of the germ f. Similarly, the values at x of all the partial derivatives of f are well--defined.

(2.3) EXERCISE

If $A,B \in P(X)$, then $(A,x)=(B,x)$ if and only if $(f_A,x)=$ $=(f_B,x)$, where f_C denotes the characteristic function of the subset C, i.e. $f_C:X \to \{0,1\}$, $f_C(a)=1$ if and only if $a \in C$.

We introduce now the classes of germs of mappings which are the most important in Singularity Theory.

(2.4) GERMS OF SMOOTH MAPPINGS

Let $X=\mathbb{R}^n$, $Y=\mathbb{R}^p$ and consider only the smooth (i.e. C^∞-differentiable) maps $f:U \to Y$, where U is some neighbourhood of a point $x \in X$. The set of all corresponding germs will be denoted by ${}_xE_{n,p}$. When $x=0$, we simply write $E_{n,p}$ for this set. Moreover, when $p=1$, i.e. when we are dealing with functions, we shall use the notations ${}_xE_n$ and, respectively, E_n.

(2.5) GERMS OF COMPLEX ANALYTIC MAPPINGS

In this case $X=\mathbb{C}^n$, $Y=\mathbb{C}^p$ and we consider only maps $f:U \to Y$ which are analytic (holomorphic), i.e. maps which can be written as a convergent power series around each point in U.

The notation ${}_xE_{n,p}$, $E_{n,p}$, ${}_xE_n$ and E_n will also be used in this context, since many properties are similar in both cases and hence can be stated in the same time. (The usual notation for E_n in Complex Analytic Geometry or Several Complex Variables books is $\mathcal{O}_n$!).

For instance, with respect to pointwise addition and multiplication, ${}_xE_n$ is a commutative unitary ring and the translation $t_x:u \to u+x$ induces a ring isomorphism $t_x^*:{}_xE_n \to E_n$. This explains why in most of the cases we consider only germs at the origin $x=0$.

We can identify the set $E_{n,p}$ with a direct product $E_n\times\ldots\times E_n$ (p factors) by associating to a map germ its components. Hence $E_{n,p}$ has a structure of a free E_n-module.

Let $E^o_{n,p}$ denote the set of germs $f \in E_{n,p}$ with $f(0)=0$. Such a germ will also be denoted by $f:(K^n,0) \to (K^p,0)$, where $K=\mathbb{R}$ or $\mathbb{C}$ according to the context. Any germ $f \in E^o_{n,p}$ induces a ring homomorphism $f^*:E_p \to E_n$ by composition: $f^*(g)=g\circ f$. Multiplication with constants in K makes E_n a K-algebra and f^* is in fact a K-algebra homomorphism.

More deep algebraic properties of the algebras E_n are discussed in the next section.

§2. BASIC PROPERTIES OF THE ALGEBRAS E_n

First we consider those properties which hold in both cases (real smooth and complex analytic).

(2.6) PROPOSITION

(i) E_n *is a local* K-*algebra and its unique maximal ideal is* $m_n=\{f\in E_n;\ f(0)=0\}$.

(ii) *The* k-*th power of the ideal* m_n *is generated by all the monomials of degree* k, *i.e. by* $x^a=x_1^{a_1}\dots x_n^{a_n}$ *with* $|a|=a_1+\dots+a_n=k$. *Moreover* $m_n^k=\{f\in E_n;\ \frac{\partial^{|a|}f}{\partial x^a}(0)=0$ *for all* a *with* $0\leq|a|<k\}$.

(iii) *An ideal* $I\subset E_n$ *has finite codimension (i.e.* $\dim_K(E_n/I)<\infty$) *if and only if* $I\supset m_n^k$ *for some integer* k.

PROOF

(i) Obviously m_n is an ideal and any element $f\notin m_n$ has an inverse $\frac{1}{f}$ which is again a germ in E_n.

(ii) We give the proof only for k=1 and leave to the reader the easy task of completing the proof by induction on k.

If $f\in m_n$, then there is a small disc $D_r=\{x\in K^n;\ |x|<r\}$ and a representative of the germ f (denoted also by f) defined on the disc D_r.

For $x\in D_r$ one has the following formula

$$(2.7)\qquad f(x)=\int_0^1 \frac{d(f(tx))}{dt}\cdot dt=\sum_{i=1,n} x_i\int_0^1 \frac{\partial f}{\partial x_i}(tx)\,dt$$

The functions $g_i(x)=\int_0^1 \frac{\partial f}{\partial x_i}(tx)dt$ define some elements in E_n and (2.7) shows that f belongs to the ideal generated by $x_1,\ldots,x_n$. To go on by induction, it is useful to notice that $g_i(0)=\frac{\partial f}{\partial x_i}(0)$.

(iii) Before starting the proof, we recall a basic result from commutative algebra which will be used quite often in the sequel.

(2.8) NAKAYAMA'S LEMMA

Let A *be a local ring,* m *its maximal ideal and* P *an* A*-module. If* $M\subset P$ *is a finitely generated submodule and* $N\subset P$ *is a submodule such that* $M\subset N+mM$*, then* $M\subset N$.

The proof, quite simple, can be found in [AM], [Lm]. A useful consequence is the following.

(2.9) COROLLARY

With the notations above, assume that P *is a finitely generated* A*-module. Let* $p_1,\ldots,p_s\in P$ *be some elements such that their classes* $\bar{p}_1,\ldots,\bar{p}_s$ *in* $P/m\cdot P$ *generate this* (A/m)*-vector space. Then* $p_1,\ldots,p_s$ *generate the* A*-module* P.

PROOF

Take in (2,8) M=P and $N=\langle p_1,\ldots,p_s\rangle$, the submodule generated by $p_1,\ldots,p_s$ in P.

Let us come back now to the proof of (2.6 iii). If $I\supset m_n^k$, then the natural epimorphism

$$\frac{E_n}{m_n^k} \to \frac{E_n}{I}$$

shows that $\dim(E_n/I)\le\dim(E_n/m_n^k)$. We can identify E_n/m_n^k with the vector space of polynomials in $x_1,\ldots,x_n$ of degree strictly less than k and this vector space is clearly finite dimensional.

Conversely, assume that $\dim(E_n/I)<\infty$. Then the infinite sequence of inclusions

$$E_n \supset I+m_n \supset I+m_n^2 \supset \dots \supset I+m_n^k \supset \dots$$

contains just a finite number of strict inclusions. Hence, there is an integer k such that $I+m_n^k=I+m_n^{k+1}$.

This clearly implies $m_n^k \subset I+m_n \cdot m_n^k$, which by Nakayama's Lemma gives $m_n^k \subset I$. Notice the use of point (ii): the ideal m_n^k is finitely generated and hence we may apply (2.8)!

(2.10) EXERCISE

Show that $\dim_K(E_n/I)=k$ implies $I \supset m_n^k$.

In order to discuss the important additional properties which hold in the complex analytic case, we start by recalling some basic properties of *germs of analytic sets*. The reader unfamiliar with these objects belonging to local Analytic Geometry should consult for instance one of the books [F], [KK], [Lm].

To any ideal $I \subset E_n$ of analytic functions corresponds a germ of a subset in $\mathbb{C}^n$ at the origin, namely

$$(2.11) \qquad V(I)=\{x \in \mathbb{C}^n;\ f(x)=0 \text{ for all } f \in I\}$$

called the *analytic (set) germ* (or the *local variety*) defined by I.

In fact, the ring E_n being noetherian (see (2.15) below), the ideal I has a finite system of generators $f_1,\dots,f_p$. And there is an open neighbourhood U of the origin on which all the germs f_j are defined (i.e. they all admit representatives defined on U). Hence the (analytic) set

$$\{x \in U;\ f_1(x)=\dots=f_p(x)=0\}$$

is well-defined and its germ at the origin is exactly V(I).

Conversely, to any germ A of a subset of $\mathbb{C}^n$ at the origin corresponds an ideal in E_n, namely $I(A)=\{f \in E_n;\ f|A=0\}$, the ideal of all function germs vanishing on A. These two correspondences $I \to V(I)$ and $A \to I(A)$ are not inverses to each other and in order to describe this phenomenon precisely, we need the following notion [AM].

If I is an ideal in some ring E, then its *radical* $\sqrt{I}$ is by definition the ideal

$$(2.12) \qquad \sqrt{I}=\{f \in E;\ f^k \in I \text{ for some positive integer } k\}.$$

Moreover, if the ring E is noetherian, then there is a positive integer p such that

$$(2.13) \qquad (\sqrt{I})^p \subset I$$

One has the following fundamental result.

(2.14) THEOREM (HILBERT NULLSTELLENSATZ)

(i) $V(I(A))=A$ *for any analytic germ* A *at the origin of* $\mathbb{C}^n$.

(ii) $I(V(J))=\sqrt{J}$ *for any ideal* $J \subset E_n$.

In fact (i) and the inclusion "$\supset$" in (ii) are simple exercises for the reader, while the remaining part of (ii) is a deep result, which can be found for instance in [Lm], p.99 where it is called Rückert's Nullstellensatz.

With these preliminaries at hand, we can pass to the following result in the complex analytic case.

(2.15) PROPOSITION

(i) *The* $\mathbb{C}$*-algebra* E_n *is isomorphic to the* $\mathbb{C}$*-algebra of convergent power series* $\mathbb{C}\{x_1,\dots,x_n\}$. *In particular,* E_n *is noetherian and factorial.*

(ii) *An ideal* $I \subset E_n$ *has finite codimension if and only if* $V(I) \subset \{0\}$.

(iii) *Any* $\mathbb{C}$*-algebra morphism* $u:E_p \to E_n$ *is induced by an analytic map germ* $g \in E^o_{n,p}$, *i.e.* $u=g^*$. *When* $n=p$, *u is an isomorphism if and only if g is an isomorphism.*

PROOF

(i) The correspondence

$$S:E_n \ni f \mapsto (\text{Taylor series of } f \text{ at the origin}) \in \mathbb{C}\{x_1,\ldots,x_n\}$$

is clearly an isomorphism. The stated properties of the convergent power series ring can be found for instance in [KK], p.80.

(ii) If I has finite codimension, then by (2.6 iii) $I \supset m_n^k$ for some k and hence

$$V(I) \subset V(m_n^k) = \{0\}$$

Conversely, assume that $V(I) \subset \{0\}$. If $V(I)=\emptyset$, then $I=E_n$ obviously.

If $V(I)=\{0\}$, then using (2.14) and the simple remark $I(\{0\})=m_n$, one gets $\sqrt{I}=m_n$. And by (2.13) we finally get $I \supset m_n^k$ for some positive integer k. By (2.6.iii) this shows that the ideal I has finite codimension.

(iii) First we show that a $\mathbb{C}$-algebra morphism $u:E_p \to E_n$ is local, i.e. $u(m_p) \subset m_n$. Assume there is a germ $f \in m_p$ such that $u(f)$ is invertible. Then $c=u(f)(0) \neq 0$ and $u(f)-c \in m_n$. Since u is a $\mathbb{C}$-algebra morphism, one has $u(c)=c$ and hence $u(f-c) \in m_n$. But $f-c$ is invertible in E_p and hence $u(f-c)$ is invertible in E_n, a contradiction.

Let now $y_1,\ldots,y_p$ be the germs in E_p corresponding to the coordinate functions and let $g_i=u(y_i) \in m_n$. In this way we get a map germ

$$g=(g_1,\ldots,g_p):(\mathbb{C}^n,0) \to (\mathbb{C}^p,0).$$

To show that $u=g^*$, note first that $u(f)=g^*(f)$ for any polynomial germ f. Using this and the obvious fact that $u(m_p^k) \subset m_n^k$, $g^*(m_p^k) \subset m_n^k$ we deduce for $f \in E_n$ that

$$u(f)-g^*(f) \in \bigcap_{k \geq 1} m_n^k = \{0\}$$

The last equality holds since an analytic function germ which vanishes at the origin together with all of its derivatives is necessarily equal to the 0 germ. Note the implicit use of (2.6.ii)! This ends the proof of the equality $u=g^*$.

Note that u induces a linear map

$$\bar{u}: m_p/m_p^2 \to m_n/m_n^2$$

which is precisely the dual of the differential $dg(0)$. This remark and the Inverse Function Theorem give the last statement in the case $n=p$.

(2.16) REMARK

In the real smooth case all the properties in (2.15) are false. More precisely, the map S constructed in the proof of (2.15 i) induces an epimorphism $E_n \to \mathbb{R}[[x_1,\ldots,x_n]]$ onto the ring of formal power series (E. Borel's Lemma, see for instance [T], p.78).

The kernel $S^{-1}(0)=m_n^\infty$ consists exactly of the flat functions and clearly:

(a) $m_n^\infty = \bigcap_{k\geq 1} m_n^k$, (b) $m_n \cdot m_n^\infty = m_n^\infty$ (use formula (2.7)!) and (c) $m_n^\infty \neq 0$.

For n=1 a classical example of flat function is $f(x)=$ $=\exp(-x^{-2})$ for $x\neq 0$ and $f(0)=0$.

By (b), (c) and Nakayama Lemma, it follows that m_n^∞ is not finitely generated and hence E_n is not noetherian.

Similarly, E_n is not factorial. For instance in E_1 , the coordinate function x to the power k divides the flat function f written above for any k(Exercise).

For the assertion (ii) consider the ideal I generated by $f=x_1^2+x_2^2$ in E_2. Then $V(f)=\{0\}$, but I has not finite codimension (Exercise).

To construct a counterexample to assertion (iii) is very subtle and the interested reader can consult [R].

§3. JETS OF MAPPINGS

Let us come back now to the map germs in $E_{n,p}^o = m_n \cdot E_{n,p}$, both real and complex cases. We shall denote by $J^k(n,p)$ the quotient vector space $E_{n,p}^o / m_n^k \cdot E_{n,p}^o$ and let $j^k: E_{n,p}^o \to J^k(n,p)$ be the canonical projection.

Obviously $J^k(n,p)$ can be identified with the K-vector

space of polynomial mappings $(K^n,0) \to (K^p,0)$, whose components have degree $\leq k$. In particular, $\dim_K J^k(n,p)<\infty$ and in fact this dimension can be explicitely computed in terms of n,p and k (Exercise).

For a map germ $f \in E^o_{n,p}$, $j^k f$ is called the k-th *jet* of f and can be thought of as being the Taylor expansion of f at the origin truncated at order k.

In particular $j^1 f$ can be identified to the differential df(0). More generally, one has the following

(2.17) EXERCISE

Show that $j^k f = j^k g$ if and only if $\frac{\partial^{|a|} f}{\partial x^a}(0) = \frac{\partial^{|a|} g}{\partial x^a}(0)$ for all multiindices a with $1 \leq |a| \leq k$.

Intuitively, the jet $j^k f$ is a finite dimensional approximation for the map germ f which is better and better as k increases.

The space $J^k(n,p)$ is called the *space of* k-*jets of type* (n,p). For s>t, there is a canonical projection $j^{s,t}: J^s(n,p) \to J^t(n,p)$ which is compatible with the projections j^s and j^t, i.e. $j^{s,t} \circ j^s = j^t$.

If we analyze now again the results in Chapter 1, we see that essentially the Submersion Theorem (resp. Morse Lemma) says that the germ of the mapping f at the origin is equal (up to a coordinate change) to the germ associated to the first jet $j^1 f$ (resp. second jet $j^2 f$) of f. The attempt to understand and generalize these special cases will be our main concern in the sequel.

Finally, we remark that for two manifolds (smooth or complex analytic) X and Y, there is a k-jet space $J^k(X,Y)$ which is a fiber bundle over $X \times Y$ with typical fiber $J^k(n,p)$, where n=dim X, p=dim Y, and which plays a basic role in studying the global mappings $f: X \to Y$. In this direction one can consult [GG].

CHAPTER 3

EQUIVALENCE RELATIONS ON GERMS AND JETS

§1. GROUP ACTIONS AND EQUIVALENCE RELATIONS

One general way to produce equivalence relations is to use group actions. All the equivalence relations introduced in this Chapter are obtained by this useful device.

More precisely, let G be a *group acting on a set* M. In other words, we are given a map

$$G\times M \to M, \qquad (g,m) \to g\cdot m$$

having the following properties

(3.1) (i) $1\cdot m=m$, for any $m\in M$, where $1\in G$ is the unit.
(ii) $g_1\cdot(g_2\cdot m)=(g_1g_2)\cdot m$, for any $m\in M$ and $g_1,g_2\in G$.

By definition, the *orbit* of an element $m\in M$ is the set of all G-translates of m, namely

$$Gm=\{g\cdot m;\ g\in G\}$$

This group action induces an *equivalence relation* $\overset{G}{\sim}$ on the set M as follows

(3.2) $m_1\overset{G}{\sim}m_2$ if and only if $Gm_1=Gm_2$.

In other words, m_2 is *equivalent* (more precisely G-*equivalent*) to m_1 if and only if m_2 is a G-translate of m_1, i.e. there exists $g\in G$ such that $m_2=g\cdot m_1$.

Here are two simple and classical examples from Linear Algebra.

(3.3) EXAMPLES

(a) Let $M=\mathrm{Hom}(K^n,K^p)$, $G=G\ell(n,K)\times G\ell(p,K)$ and the multiplication rule $(g,h)\cdot u=hug^{-1}$ for $u\in M$, $(g,h)\in G$. It is easy to show that $u_1\overset{G}{\sim}u_2$ if and only if the linear mappings u_1 and u_2

have the same rank.

Hence any linear map $u \in M$ is equivalent to a *normal form* $u_k \in M$

$$u_k(x_1,\ldots,x_n)=(x_1,\ldots,x_k,\ 0,\ldots,0)$$

where $k=\text{rank } u$. In particular, there are exactly $\min(n,p)+1$ orbits.

(b) Let $M=\text{Hom}(\mathbb{C}^n,\mathbb{C}^n)$, $G=G\ell(n,\mathbb{C})$ and the multiplication rule $g\cdot u=gug^{-1}$ for $g\in G$, $u\in M$. Then $u_1 \overset{G}{\sim} u_2$ if and only if the linear endomorphisms u_1 and u_2 (or the associated square matrices) have the same *Jordan normal form*. In particular, for $n\geqslant 1$ there are infinitely (uncountable) many orbits.

In general, for a group G acting on a set M, the subgroup of elements in G which fix an element $m\in M$, namely

$$G_m=\{g\in G;\ g\cdot m=m\}$$

is called the *isotropy subgroup of the element* m.

The groups which are important in the study of map germs are constructed from the following two types of groups.

We shall denote by D_n the group of map germs $g:(K^n,0)\to (K^n,0)$ which are diffeomorphisms for $K=\mathbb{R}$ and analytic isomorphisms for $K=\mathbb{C}$, the group operation being the composition of map germs. Note that in both cases a germ $g\in E^o_{n,n}$ belongs to D_n if and only if its differential $dg(0)$ is a linear isomorphism.

We shall denote by $M_{n,p}$ the general linear group $G\ell(p,E_n)$, i.e. the group of invertible square matrices of order p with entries in the K-algebra E_n of function germs. There is a well-defined evaluation homomorphism $G\ell(p,E_n)\to G\ell(p,\mathbb{C})$, $A\to A(0)$.

The following equivalence relations on the space of map germs $E^o_{n,p}$ are standard [W2].

(3.4) RIGHT EQUIVALENCE (R-EQUIVALENCE)

This equivalence relation is associated to the action $r:D_n\times E^o_{n,p}\to E^o_{n,p}$, $(g,f)\to f\circ g^{-1}$.

As an example, note that in (1.1) and (1.3) we actually get the R-equivalence of a germ f with a corresponding normal form.

Note that we can define an action (and hence a corresponding equivalence relation) on the larger space $E_{n,p}$ by the same formula as above. There is a natural projection

$$E_{n,p} \to E^o_{n,p}\ ,\quad f \to f_o=f-f(0)$$

and, for $f,g \in E_{n,p}$, one has $f \sim g$ if and only if $f_o \sim g_o$, $f(0)=g(0)$.

Therefore it is enough to study the action r and the corresponding equivalence relation on the space $E^o_{n,p}$.

(3.5) LEFT EQUIVALENCE (*L*-EQUIVALENCE)

This equivalence relation comes from the action

$$l: D_p \times E^o_{n,p} \to E^o_{n,p}\ ,\quad (h,f) \to h \circ f\ .$$

A simple example where this occurs is the analogue of Theorem (1.1) in the case of immersions:

An immersion germ $f \in E^o_{n,p}$ *is L-equivalent to the germ of the linear immersion*

$$(x_1,\dots,x_n) \to (x_1,\dots,x_n,\ 0,\dots,0).$$

We recall that $f \in E^o_{n,p}$ is called an *immersion* (germ) if rank $df(0)=n$.

(3.6) RIGHT-LEFT EQUIVALENCE (*A*-EQUIVALENCE)

This equivalence relation is associated to the action

$$a: (D_n \times D_p) \times E^o_{n,p} \to E^o_{n,p}\ ,\quad (g,h)\cdot f=h \circ f \circ g^{-1}\ .$$

Hence the group here is the *direct product* of the groups D_n and D_p and, in fact, the action a is obtained by putting together the actions r and l introduced above.

It is usual to represent two *A*-equivalent germs f_1 and f_2 by a commutative diagram

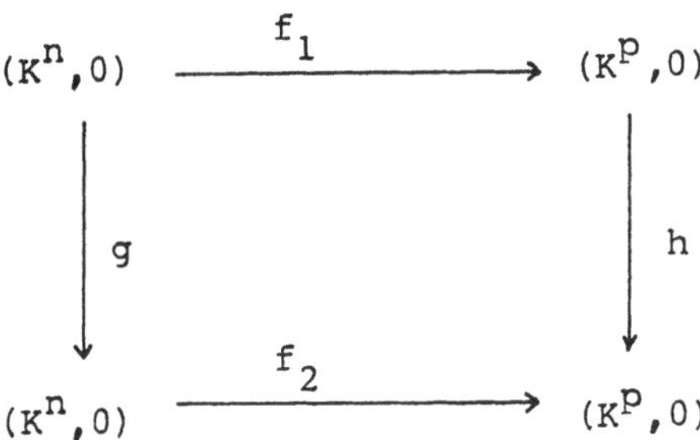

where $f_2=(g,h)\cdot f_1$. In more down-to-earth terms, the germs f_1 and f_2 coincide in suitable coordinate systems at the source and at the target.

As an example where A-equivalence occurs, we offer the next result containing as special cases the Submersion Theorem (1.1) and the Immersion Theorem just stated above.

(3.7) CONSTANT RANK THEOREM

Let U *be a neighbourhood of* $0\in K^n$ *and* $f:U\to K^p$ *be a smooth (for* K=R) *resp. analytic (for* K=ℂ) *mapping. Assume that* f(0)=0 *and that the rank of the differential of* f *is constant on* U*, let's say*

$$\operatorname{rank} df(x)=m\leq\min(n,p), \quad \textit{for all} \quad x\in U.$$

Then the germ of f *in* $E^o_{n,p}$ *is A-equivalent to the germ of the linear mapping*

$$(x_1,\ldots,x_n)\longmapsto(x_1,\ldots,x_m\ ,\ 0,\ldots,0).$$

PROOF

Replacing U with a smaller neighbourhood V of 0 and re-indexing the coordinates, we can assume that

$$\operatorname{rank}\left|\frac{\partial f_i}{\partial x_j}(x)\right|_{i,j=1,\ldots,m}=m$$

for any $x \in V$.

Let $q: K^p \to K^m$ denote the projection on the first m coordinates. Then $q \circ f$ is a submersion and by (1.1) there is a diffeomorphism g such that $q \circ f \circ g^{-1}$ is equal to the linear projection

$$(x_1, \ldots, x_n) \longmapsto (x_1, \ldots, x_m) .$$

It follows that the map $f \circ g^{-1}$ is given by $x \longmapsto (x_1, \ldots, x_m, \bar{f}_{m+1}(x), \ldots, \bar{f}_p(x))$ for some functions $\bar{f}_j$ (which are defined on $\mathbb{R}^n$ in the real case and on an open ball B in the complex case - the reader is asked to restate carefully (1.1) in the complex analytic case! The idea is that B is not isomorphic to $\mathbb{C}^n$).

The condition of constant rank on f (and hence on $f \circ g^{-1}$) implies that

$$\frac{\partial \bar{f}_j}{\partial x_k}(x) = 0 \text{ for any } j = m+1, \ldots, p;\ k = m+1, \ldots, n$$

and any x. Hence the functions $\bar{f}_j$ depend only on $x_1, \ldots, x_m$. We consider them as germs in E_m and define an element $h \in D_p$ by the formula

$$h(y) = (y_1, \ldots, y_m, y_{m+1} - \bar{f}_{m+1}(\bar{y}), \ldots, y_p - \bar{f}_p(\bar{y}))$$

where $\bar{y} = (y_1, \ldots, y_m)$.

Then clearly $h \circ f \circ g^{-1}$ coincides with the linear mapping in the statement of (3.7).

(3.8) CONTACT EQUIVALENCE ($\mathcal{K}$-EQUIVALENCE)

Let G be the *semidirect product* $D_n \rtimes M_{n,p}$ given by the multiplication rule

$$(h_1, A_1) \cdot (h_2, A_2) = (h_1 h_2,\ A_1 \cdot (A_2 \circ h_1^{-1}))$$

where composition with h_1^{-1} refers to all the entries of the matrix A_2. This group G will be also denoted by $\mathcal{K}_{n,p}$ or simply by $\mathcal{K}$.

The K-equivalence is associated to the action

$$k:G\times E^{o}_{n,p}\to E^{o}_{n,p}\ ,\quad (h,A)\cdot f=A(f\circ h^{-1})$$

where the germs f and $f\circ h^{-1}$ should be considered as column vectors with entries in the maximal ideal $m_n\subset E_n$. The multiplication in the right-hand side is then matrix multiplication.

A geometric explanation for the use of the word *contact* here can be found in [GG], p. 172 or in our book on p. 215-216.

This contact equivalence relation has a basic rôle in the *classification of stable map germs*, a fundamental topic in Singularity Theory (e.g. [Gi], [Mr]) which we do not touch at all in our book.

There is another reason which makes K-equivalence a basic tool in many parts of Algebraic and Analytic Geometry, namely its connection to the germs of analytic spaces. This connection will be considered in detail in the next section.

§2. GERMS OF COMPLEX ANALYTIC SPACES

In this section we shall briefly introduce the (complex) analytic space germs and some related notions. A more extensive presentation can be found in [Lm], Chap. III, while for a complete treatment we send to [KK].

Recall that to any ideal $I\subset E_n$ of complex analytic functions germs corresponds a germ of analytic set V(I) at the origin 0 of $\mathbb{C}^n$. By Hilbert Nullstellensatz (2.14) we learn that this set germ V(I) does not determine completely the ideal I, but only its radical $\sqrt{I}$.

To avoid this loss of symmetry between ideals in E_n and germs of analytic sets at 0 in $\mathbb{C}^n$, in local Analytic Geometry one introduces the following objects.

(3.9) DEFINITION

The *analytic space germ* defined by the ideal $I\subset E_n$ is a pair $(X, \mathcal{O}_X)$, where X=V(I) and $\mathcal{O}_X$ is the local $\mathbb{C}$-algebra E_n/I.

Usually we shall denote such a germ simply by X, omitting the algebra $\mathcal{O}_X$.

For instance, the notation

$$X: f_1 = \ldots = f_p = 0, \quad \text{where} \quad f_i \in m_n \subset E_n$$

will define an analytic space germ corresponding to the ideal $I=(f_1,\ldots,f_p)$ generated by the *equations* f_i in E_n. The same analytic space germ X is sometimes called the *fiber* $f^{-1}(0)$ of the map germ $f=(f_1,\ldots,f_p):(\mathbb{C}^n,0) \to (\mathbb{C}^p,0)$.

For basic notions on analytic space germs (e.g. dimension, irreducible components) we send to the above references.

An analytic space germ $(X,\mathcal{O}_X)$ as above is sometimes regarded as a subgerm in $(\mathbb{C}^n,0)$ (notation $(X,0)\subset(\mathbb{C}^n,0)$) and its *codimension* is then

$$\operatorname{codim} X = n - \dim X.$$

Two germs of analytic spaces $(X,\mathcal{O}_X)$ and $(X',\mathcal{O}_{X'})$ are called *isomorphic* if the corresponding $\mathbb{C}$-algebras $\mathcal{O}_X$ and $\mathcal{O}_{X'}$ are isomorphic (notation $X\sim X'$.) If $(X,\mathcal{O}_X)$ is an analytic space germ, then its *tangent vector space* is defined by the equality

(3.10) $\quad TX = \operatorname{Hom}\,(m_X/m_X^2,\ \mathbb{C})$

where m_X is the maximal ideal in $\mathcal{O}_X$.

The number dim TX is called the *embedding dimension* of X and is denoted by edim X.

(3.11) EXERCISE

(i) Assume that the analytic space germ X is the fiber of the map germ $f \in E^o_{n,p}$. Then

$$\operatorname{edim} X = n - \operatorname{rank} df(0).$$

(ii) Show that two isomorphic analytic space germs have the same embedding dimension.

(3.12) DEFINITION

An analytic space germ $(X,0)\subset(\mathbb{C}^n,0)$ and the corresponding ideal $I_X \subset m_n$ are called *minimally embedded* if edim X=n.

(3.13) PROPOSITION

(i) $(X,0) \subset (\mathbb{C}^n,0)$ *is minimally embedded if and only if* $I_X \subset m_n^2$.

(ii) *For any analytic space germ* X, *there is an analytic space germ* $(Y,0) \subset (\mathbb{C}^k,0)$ *which is minimally embedded and such that* X *is isomorphic to* Y. *Moreover* k=edim X.

We can rephrase (ii) by saying that k=edim X is the minimal dimension of a smooth germ in which X can be embedded.

PROOF

For (i) note that if I_X is a proper ideal (i.e. $I_X \neq E_n$) then $m_X/m_X^2 \simeq m_n/I+m_n^2$.

(ii) Let k=edim X and $f_1,\dots,f_p$ be a set of generators for the ideal I_X, $f=(f_1,\dots,f_p) \in E^o_{n,p}$ the corresponding map germ.

By (3.11.i) it follows that rank df(0)=n-k. An argument similar to that in the proof of (3.7) shows that the germ X is isomorphic to an analytic space germ X' in $\mathbb{C}^n$ defined by a system of equations of the form

$$x_1=\dots=x_{n-k}=\bar{f}_{n-k+1}(x)=\dots=\bar{f}_p(x)=0$$

where $g_j(\bar{x})=\bar{f}_{n-k+j}(0,\bar{x}) \in m_k^2$, $\bar{x}=(x_{n-k+1},\dots,x_n)$ and $j=1,\dots,p-n+k$.

If we denote by Y the analytic space germ defined in $\mathbb{C}^k$ by the equations $g_j=0$, then the above formulas show that $X \sim Y$ and that Y is minimally embedded.

Therefore it is enough to consider, most of the time, minimally embedded analytic space germs and this may simplify the proofs (see for instance (3.16)).

We give now a formal answer to a question which maybe the reader has put to himself from the very beginning of this book: what is a singularity?

(3.14) DEFINITION

(i) A map germ $f \in E^o_{n,p}$ (both real and complex cases) or its R-, L-, A- or K-equivalence class is called a *singularity*. (One usually assumes in this case that

$$\text{rank } df(0) < \min(n,p)$$

i.e. one avoids the specially simple cases of immersions and submersions.)

(ii) An analytic space germ $(X, \mathcal{O}_X)$ or its isomorphism class is called a *singularity*.
(One usually assumes in this case that the local ring $\mathcal{O}_X$ is not regular, i.e. $(X,0)$ is not isomorphic to the smooth germ $(\mathbb{C}^m, 0)$ for some $m \geq 0$.)

(3.15) EXERCISE

Show that an analytic space germ $(X, \mathcal{O}_X)$ is smooth (i.e. $\mathcal{O}_X$ is regular) if and only if dim X=edim X. If necessary use results on regular rings from [AM], Chap. 11.

The connection between the two concepts of singularity above (in the complex case) is the following.

Represent X as the fiber of a map germ $f \in E^o_{n,p}$. Then X is a singularity (i.e. $\mathcal{O}_X$ is not regular) if and only if rank $df(0) < \text{codim } X$. This is a simple application of Implicit Functions Theorem, see for instance [Mu], p. 9 or use (3.15). Compare also to (6.37 ii) below.

Now we shall describe the connection between K-equivalence and (isomorphisms of) germs of analytic spaces as announced at the end of the previous section.

An analytic space germ $(X, \mathcal{O}_X)$ is called *reduced* if the ring $\mathcal{O}_X$ has no nilpotent elements (i.e. there is no element $a \in \mathcal{O}_X$, $a \neq 0$ and $a^m = 0$ for some $m \geq 0$).

This is clearly equivalent to $I_X = \sqrt{I_X}$, where I_X is the ideal defining the germ X. Hence, by Hilbert Nullstellensatz (2.14), in the case of a reduced analytic space germ, the ideal I_X (and the algebra $\mathcal{O}_X$) is determined by the set germ X. We shall identify a reduced analytic space germ with the corresponding analytic set germ. And an analytic set germ $(X,0) \subset (\mathbb{C}^n, 0)$ will be regarded as a reduced analytic space germ with the corresponding local algebra $\mathcal{O}_X = E_n/I(X)$, where

$I(X)$ is the ideal of all function germs vanishing on X.

The converse is always true: the local algebra $\mathcal{O}_X$ determines the analytic set germ X (this was implicit in the definition of isomorphic analytic space germs above!)

In fact, the correspondence $(X,\mathcal{O}_X) \to \mathcal{O}_X$ sets up *an anti--equivalence from the category of analytic space germs to the category of analytic* $\mathbb{C}$*-algebras*, see for instance [F], p. 16 for the definition of morphisms in these categories and for a proof.

Before stating the result we are looking for, we need some more definitions. Consider two germs of analytic spaces defined both by p equations at the origin $0 \in \mathbb{C}^n$

$$X: f_1 = \ldots = f_p = 0 \quad \text{and} \quad X': f_1' = \ldots = f_p' = 0$$

(where $f_i,\ f_j' \in m_n$).

Let $I_X = (f_1,\ldots,f_p)$ and $I_{X'} = (f_1',\ldots,f_p')$ be the corresponding ideals in E_n and let $f,\ f' \in E^o_{n,p}$ be the associated map germs.

We say that the germs of analytic spaces $(X,\mathcal{O}_X)$ and $(X',\mathcal{O}_{X'})$ are:

(a) *ambient isomorphic*, if there is an element $g \in D_n$ such that $g^*(I_{X'}) = I_X$.

(b) *geometric ambient isomorphic*, if there is an element $g \in D_n$ such that $g(X) = X'$ (equality of set germs).

With these preliminaries, we have the following basic result.

(3.16) PROPOSITION

Consider the following statements:

(i) $(X,\mathcal{O}_X)$ *and* $(X',\mathcal{O}_{X'})$ *are isomorphic.*

(ii) $(X,\mathcal{O}_X)$ *and* $(X',\mathcal{O}_{X'})$ *are ambient isomorphic.*

(iii) $(X,\mathcal{O}_X)$ *and* $(X',\mathcal{O}_{X'})$ *are geometric ambient isomorphic.*

(iv) *The map germs* f *and* f' *are* K*-equivalent. Then* (i) $\Longleftrightarrow$ (ii) $\Longleftrightarrow$ (iv) $\Longrightarrow$ (iii). *Moreover, if the analytic space germs* X *and* X' *are both reduced, then* (iii) $\Longrightarrow$ (ii).

PROOF

The implications (iv)$\Longrightarrow$(ii)$\Longrightarrow$(i) and (ii)$\Longrightarrow$(iii) are obvious.

We prove first that (i)$\Longrightarrow$(ii). We treat only the special case when X and X' are minimally embedded and leave to the reader the easy task to complete the proof in the general case.

Let $u: \mathcal{O}_{X'} \to \mathcal{O}_X$ be an isomorphism of $\mathbb{C}$-algebras and let $g_1, \ldots, g_n \in m_n$ be germs such that

$$g_i \bmod I_X = u(x_i \bmod I_{X'}) \qquad \text{for } i=1,\ldots,n.$$

We get thus a map germ $g=(g_1,\ldots,g_n) \in E^o_{n,n}$. The induced morphism $g^*: E_n \to E_n$ has the property $g^*(I_{X'})=I_X$. Indeed, if $h \in I_{X'}$ then $g^*(h) \bmod I_X = h(g_1,\ldots,g_n) \bmod I_X = u\ (h \bmod I_{X'}) =$ $= u(0)=0$.

(Recall the proof of(2.15 iii)if necessary!)

To show that $g \in D_n$, note that the dual of the differential $dg(0)$ can be identified to the linear isomorphism

$$m_n/m_n^2 \simeq m_{X'}/m_{X'}^2 \xrightarrow{\bar{u}} m_X/m_X^2 \simeq m_n/m_n^2$$

induced by u.

Let us prove now that (ii)$\Longrightarrow$(iv). By (ii) we get two p×p matricesA, B with entries in E_n such that

$$f' = A \cdot \bar{f} \text{ and } \bar{f} = B \cdot f', \text{ where } \bar{f} = f \circ g^{-1}.$$

for some $g \in D_n$ and f, f', $\bar{f}$ are considered as column vectors.

The only problem is that the matrices A and B are not necessarily invertible. To repair this, we need the following elementary result from Linear Algebra.

(3.17) LEMMA

Let G *and* H *be* p×p *matrices over a field* k. *Then there is a matrix* F *of the same type such that*

$$L = F(1-GH)+H$$

is an invertible matrix.

PROOF OF THE LEMMA

We identify a p×p matrix over k to the corresponding linear endomorphism of the vector space k^p.

Choose subspaces V and W in k^p such that $V \oplus \ker H = W \oplus \operatorname{im} H = k^p$, where $\oplus$ stands for direct sum. Next take an endomorphism F of k^p such that $F|V=0$ and F induces an isomorphism $\ker H \to W$. With this choice for F, the endomorphism L defined in the statement is invertible.

Indeed, it is enough to show that ker L=0. If Lv=0, then

$$\operatorname{im} H \ni Hv=-F(1-GH)v \in W$$

and hence $v \in \ker H$, F(v)=0. And this is possible only for v=0.

To finish the proof of the implication (ii)$\Longrightarrow$(iv) let us put G=A(0), H=B(0) and define a new matrix with entries in E_n

$$\bar{L}=F(1-AB)+B$$

where F is the matrix found in (3.17).

Then $\bar{L}(0)=L$ is invertible by (3.17) and hence $L \in M_{n,p}$. The obvious equality $\bar{L}\cdot f'=\bar{f}$ shows that f is K-equivalent to f'.

Finally we prove that (iii)$\Longrightarrow$(ii) under the assumption that the germs X and X' are reduced. By Hilbert Nullstellensatz (2.14) one has $I_X=I(X)$, $I_{X'}=I(X')$. But g(X)=X' obviously implies $g^*(I(X'))=I(X)$ and this ends the proof.

As a consequence, we prove the following implications among the equivalence relations introduced in the previous section. First some notations: we let Rf, Lf, Af and Kf denote the corresponding orbits of a map f, $f \overset{R}{\sim} g$ mean f is R-equivalent to g (similarly for L, A, K) and $R \Longrightarrow A$ mean that for any germs f and g such that $f \overset{R}{\sim} g$ one has $f \overset{A}{\sim} g$ (in other words: $Rf \subset Af$).

(3.18) COROLLARY

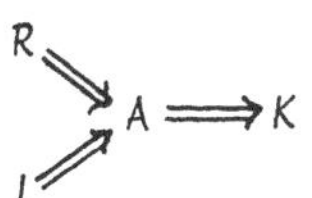

PROOF

The implications $R \Longrightarrow A$, $L \Longrightarrow A$ are obvious from the definitions.

To prove $A \Longrightarrow K$ *in the complex case* we proceed as follows. To a map germ $f \in E^o_{n,p}$ we associate the ideal I_f generated in E_n by its components. Note that I_f is also the ideal in E_n generated by $f^*(m_p)$.

By the equivalence (ii)$\Longleftrightarrow$(iv) in (3.16) we know that $f \overset{K}{\sim} g$ if and only if there is an isomorphism $h \in D_n$ such that $h^*(I_g)=I_f$.

If $f \overset{A}{\sim} g$, then $g=h_1 \circ f \circ h_2^{-1}$ for some isomorphisms $h_1 \in D_p$, $h_2 \in D_n$. One has

$$g^*(m_p)=(h_2^*)^{-1} \circ f^* \circ h_1^*(m_p)=(h_2^*)^{-1} \circ f^*(m_p) \quad .$$

Hence $h_2^*(I_g)=I_f$ and this ends the proof .

(3.19) EXERCISE

(i) Show that the equivalence (ii)$\Longleftrightarrow$(iv) in (3.16) holds essentially in the real case too. Use this fact to complete the proof of the implication $A \Longrightarrow K$ above.

(ii) Show by explicit examples that no implication in (3.18) can be reversed.

Before ending this section, we point out a useful *enlargement* of the category of map germs to which the notions and the results presented in this book apply. Let M, N be two manifolds (real smooth or complex analytic) and let $x \in M$, $y \in N$ be two fixed points. A smooth (or complex analytic) map $f:M \to N$ with $f(x)=y$ induces a map germ $f_x:(M,x) \to (N,y)$. Choose parametrizations $h:(K^m,0) \to (M,x)$ and $k:(K^n,0) \to (N,y)$, where $m=\dim M$, $n=\dim N$. We get thus a map germ $\bar{f}_x=k^{-1} \circ f_x \circ h$ in $E^o_{m,n}$ and moreover its A-equivalence class (and hence its K-equivalence class by (3.18)) does not depend on the choice of the parametrizations.

This well-defined A or K equivalence class is called the A (resp. K) equivalence class of the map germ f_x.

A similar remark holds for the R equivalence class of a function germ $f_x:(M,x) \to (K,0)$.

§3. GROUP ACTIONS ON JET SPACES

We come back now to the main theme of this Chapter, namely the construction of equivalence relations via group actions.

If we compare the actions from Example (3.3) to the actions from the definitions (3.4), (3.5), (3.6) and (3.8), we notice the following basic difference.

The first actions are *algebraic*, i.e. G is an algebraic group (real or complex), the set M on which G acts is a smooth algebraic variety (even a finite dimensional vector space) and the multiplication map $G\times M \to M$ is a morphism of algebraic varieties. The algebraic actions on a smooth variety are special cases of smooth actions and hence we have at our disposal a lot of tools for their study.

The reader may find useful to have a look at the first sections of Humphreys' book [H2] for basic results on algebraic groups.

On the other hand, in the definitions of the $\mathcal{R}$, $\mathcal{L}$, $\mathcal{A}$ and $\mathcal{K}$ equivalence relations, the set M is the infinite dimensional vector space $E^0_{n,p}$ and the corresponding groups G are themselves infinite dimensional objects.

In the end of the previous Chapter we introduced the jet spaces $J^k(n,p)$ as finite dimensional approximations for the space $E^0_{n,p}$. Just in the same way we construct now some finite dimensional approximations (i.e. some algebraic actions) corresponding to the actions in (3.4), (3.5), (3.6) and (3.8).

By its definition, $D_n \subset E^0_{n,n}$ and hence we can define $D^k_n = j^k(D_n) \subset J^k(n,n)$ for any $k \geq 1$. Since $j^k f \in D^k_n$ if and only if $df(0) = j^1 f$ is a linear isomorphism, it follows D^k_n is a Zariski open set in $J^k(n,n)$ and, more precisely, it is the complement of a hypersurface.

Hence D^k_n is a smooth affine algebraic variety. The set $D_n(k) = \{g \in D_n;\ j^k g = j^k 1\}$, where $1 \in D_n$ denotes the identity, is a normal subgroup in D_n and for $g_1, g_2 \in D_n$ one has

$$j^k g_1 = j^k g_2 \iff g_1 g_2^{-1} \in D_n(k)$$

It follows that the quotient group $D_n/D_n(k)$ can be identified with the variety D_n^k.

The resulting *algebraic group* is denoted again by D_n^k and is called the *group of* k-*jets of diffeomorphisms* (in both real and complex cases!). As the simplest example, note that $D_n^1=G\ell(n,K)$.

To make a similar construction for the matrix group $M_{n,p}$ note first that for $A\in M_{n,p}$ the matrix $A-A(0)$ has all its entries in the maximal ideal m_n.

For any positive integer $k\geq 1$ the set

$$M_{n,p}(k)=\{A\in M_{n,p};\ A(0)=1,\ j^{k-1}(A-1)=0\}\ ,$$

where 1 is the unit matrix in $G\ell(p,K)$, is a normal subgroup in $M_{n,p}$.

It is easy to show that the quotient group $M_{n,p}^k=$ $=M_{n,p}/M_{n,p}(k)$ is an algebraic group isomorphic to the general linear group $G\ell(p,\ E_n/m_n^k)$.

For any positive integer $t\geq 1$, we define the following groups and group actions:

$$\text{(3.20)}\qquad R_n^t=D_n^t\ ,\quad r^t:R_n^t\times J^t(n,p)\to J^t(n,p)$$
$$(j^tg,j^tf)\longmapsto j^t(f\circ g^{-1}).$$

$$\text{(3.21)}\qquad L_p^t=D_p^t\ ,\quad \ell^t:L_p^t\times J^t(n,p)\to J^t(n,p)$$
$$(j^th,j^tf)\longmapsto j^t(h\circ f).$$

$$\text{(3.22)}\qquad A_{n,p}^t=D_n^t\times D_p^t\ ,\quad a^t:A_{n,p}^t\times J^t(n,p)\to J^t(n,p)$$
$$((j^tg,j^th),j^tf)\longmapsto j^t(h\circ f\circ g^{-1}).$$

$$\text{(3.23)}\qquad K_{n,p}^t=D_n^t\ltimes M_{n,p}^t\ ,\quad k^t:K_{n,p}^t\times J^t(n,p)\to J^t(n,p)$$
$$((j^tg,j^{t-1}A),\ j^tf)\longmapsto j^{t-1}A\cdot j^t(f\circ g^{-1}),$$

where $\ltimes$ stands for a semidirect product as in (3.8) and $j^{t-1}A$ denotes the class of a matrix $A\in M_{n,p}$ in the truncated group

$M^t_{n,p}$.

It is easy to check that all these actions are well-defined and algebraic. Moreover the diagram

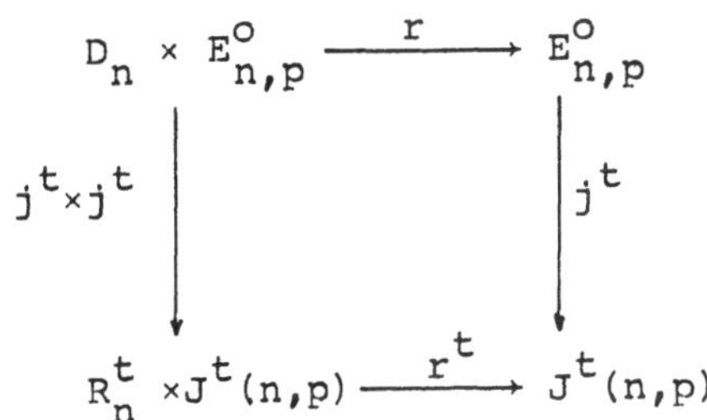

and all the other similar diagrams are commutative. This is the exact sense in which the actions (3.20)-(3.23) are approximations for the actions r, ℓ, a and k.

The equivalence relations on the jet spaces $J^t(n,p)$ induced by these actions are also called R, L, A and respectively K equivalence. Note that the action in Example (3.3 a) coincides to the actions a^1 and k^1 with our present notations.

In the next Chapter we start a detailed study of the above actions on jet spaces.

CHAPTER 4

TANGENT SPACES TO ORBITS

§1. SEMIALGEBRAIC SETS

We recall here some basic facts on semialgebraic sets, which are also called *constructible sets*, especially in the complex case. For a more complete introduction to this subject one can see [GWPL], Chap. 1 and for complete proofs in a more general setting one can consult [Ło].

Let M be a smooth algebraic variety over K ($K = \mathbb{R}$ or $\mathbb{C}$ as usual). In most of our applications, M is just an affine space K^m.

(4.1) DEFINITION

(i) COMPLEX CASE. A subset $A \subset M$ is called *semialgebraic* if A belongs to the Boolean subalgebra generated by the Zariski closed subsets of M in the Boolean algebra $P(M)$ of all subsets of M.

(ii) REAL CASE. A subset $A \subset M$ is called *semialgebraic* if A belongs to the Boolean subalgebra generated by the open sets

$$U_f = \{x \in U;\ f(x) > 0\}$$

where $U \subset M$ is an algebraic open subset in M and $f: M \to \mathbb{R}$ is an algebraic function, in the Boolean algebra $P(M)$ of all subsets of M.

By definition, it follows that the class of semialgebraic subsets of M is closed under finite unions, finite intersections and complements. If $f: M \to N$ is an algebraic mapping among the smooth algebraic varieties M and N and if $B \subset N$ is semialgebraic, then cleary $f^{-1}(B)$ is semialgebraic in M. Conversely, one has the following basic result.

(4.2) PROPOSITION (TARSKI-SEIDENBERG-CHEVALLEY)

If $A \subset M$ *is semialgebraic, then* $f(A) \subset N$ *is also semialgebraic.*

(4.3) EXERCISE

Let $f:K^2 \to K^2$ be the map given by $f(x,y)=(x,xy)$. Determine $f(K^2)$ and show that this semialgebraic set is not locally closed (neither in the Zariski nor in the strong topology of K^2).

Let $A \subset M$ be a semialgebraic set. A point $x \in A$ is *regular (of dimension* d) if x has a neighbourhood U in M such that $A \cap U$ is a smooth submanifold of U (of dimension d over K). The *dimension* of A is the maximal dimension of regular points in A and this is well-defined since the regular points always exist if A is nonempty. The points in A which fail to be regular of maximal dimension are called *singular*. The *singular set* S(A) of A is the set of all such singular points in A. It is easy to see that S(A) is closed in A. When $S(A)=\emptyset$, we say that A is *smooth*, especially when A is irreducible.

It is natural to compare this notion of singular points with our Definition (3.14). In the complex case, they essentially coincide:

(4.4) PROPOSITION (see [M1], p.13)

Let $X \subset M$ *be a closed semialgebraic set and* $x_0 \in X$ *be a point such that* $\dim_{x_0} X = \dim X$. *Then* x_0 *is a regular point if and only if* (X,x_0) *is a smooth germ* i.e. *the corresponding local analytic ring* O_X *is regular.*

PROOF

Note first that the germ (X,x_0) of the semialgebraic set X is indeed an analytic germ. By $\dim_{x_0} X$ we mean precisely the dimension of this germ.

The problem being local on M, we can assume that M is an open subset in some $\mathbb{C}^n$ and that $x_0=0$. If $(X,0)$ is a smooth germ (with $O_X=E_n/I(X)$) then it follows, by (3.15) and the comments just after it, that 0 is a regular point in the above sense.

Conversely, assume that 0 has a neighbourhood in X which is (only!) a C^1-submanifold of $\mathbb{C}^n$ of real dimension 2p (this differentiable structure has to coincide with that induced by the complex structure at the smooth points on X).

If $x \in X$ is such a smooth point, then the real tangent space $T_xX \subset T_x\mathbb{C}^n \simeq \mathbb{C}^n$ is actually a complex subspace: $i(T_xX)=T_xX$. Since the smooth points are dense in X, it follows by continuity that the same is true for all tangent spaces T_xX.

Renumbering the coordinates if necessary, the Implicit Function Theorem shows that the germ (X,0) coincides to the germ at the origin of the graph of a C^1-map germ $g:(\mathbb{C}^p,0)\to(\mathbb{C}^{n-p},0)$ This identification and the above property of the tangent spaces T_xX show that the germ g has a $\mathbb{C}$-linear differential dg(u) for any $u \in \mathbb{C}^p$, sufficiently close to the origin. In other words, the Cauchy-Riemann equations are satisfied by g and hence g is complex analytic [KK], p.4.

It follows that X=graph(g) is a smooth submanifold germ in $\mathbb{C}^n$ and hence the local ring $\mathcal{O}_X$ is regular.

(4.5) COROLLARY

A complex analytic space germ (X,0) *is singular if and only if one of the following two cases occur: either the germ* (X,0) *is not reduced or the germ* (X,0) *is reduced, but it is not a smooth manifold around* 0.

As usual, in the real case, the connections between geometry (point sets) and algebra (functions) are more loose.

(4.6) EXAMPLE (see [M1], p.12)

Consider the real plane curve C given by the equation $F(x,y)=y^3+2x^2y-x^4=0$ in $\mathbb{R}^2$. Noting that $(x,y)\in C$ implies $y\geq 0$, one can solve for x^2 and get the following

$$(x,y)\in C \iff f(x,y)=0, \quad \text{where} \quad f(x,y)=x^2-y(1+\sqrt{1+y})$$

Next, the origin 0=(0,0) is a *singular point for* F (dF(0)=0) and F is an irreducible polynomial equation for C.

Our computations above show that $f=0$ can be taken as a local real analytic equation for C at the origin. Since $df(0)\neq 0$, it follows that C is a smooth (even real analytic) manifold at the origin, i.e. 0 is a *regular point for* C in the above sense.

A basic result which holds in both real and complex cases is the following.

(4.7) PROPOSITION

For any semialgebraic set A, *the singular set* S(A) *is also semialgebraic with* $\dim S(A)<\dim A$. *In particular* $A\neq S(A)$.

With these preliminaries, we can prove the following useful result.

(4.8) PROPOSITION

Let G *be an algebraic group acting (algebraically) on a smooth algebraic variety* M. *Then the corresponding orbits are smooth semialgebraic subsets in* M.

PROOF

Let $x\in M$ be a point and $X=G\cdot x$ be its orbit. Then X is the image of the morphism of algebraic varieties $G\to M$, $g\longmapsto g\cdot x$ and hence X is semialgebraic by (4.2).

By (4.7) there is a point $y\in X\setminus S(X)$. Since $y\in X$ one has $y=g\cdot x$ for some $g\in G$. Multiplication by g^{-1} induces an isomorphism of the pair $(U, U\cap X)$ onto the pair $(g^{-1}(U), g^{-1}(U)\cap X)$, where U is any neighbourhood of y in M. Since $g^{-1}(U)$ is then a neighbourhood of x, it follows that $x\in X\setminus S(X)$ as well.

(4.9) REMARK

With the notations above, the closure of the orbit X is again a semialgebraic set with $\dim \overline{X}=\dim X$, (see [GWPL] p.18-19) and $\overline{X}\setminus X$ is a union of G-orbits of strictly less dimension than dim X.

Note that by the quoted result, the closures of a semialgebraic set in the Zariski topology and in the strong (metric) topology coincide in the *complex case* (i.e. a closed semialgebraic set is the zero set of some polynomials). In the real case this is no longer true (e.g. consider a closed ball).

Show by an example that $X \neq \operatorname{int} \bar{X}$ in general, where int denotes the interior.

§2. TANGENT SPACES AND TRANSVERSALS

Let $m: G \times M \to M$ be an algebraic action and $X = G \cdot x$ the orbit of a point $x \in M$. Then by (4.8) X is a smooth submanifold of M and hence we can consider the tangent space $T_x X \subset T_x M$. One has obviously

(4.10) $\quad T_x X = \operatorname{im}\, dm_x(e)$

where $m_x: G \to M$, $m_x(g) = g \cdot x$ and e denotes the unit element in the group G.

Let $T_e G$ denote the *Lie algebra* of the group G, identified to the tangent space to G at e and let $\exp: T_e G \to G$ denote the *exponential map* [Ad], p. 10. Then one has

(4.11) $\quad T_x X = \{\frac{d}{dt}[\exp(tv) \cdot x]_{t=0} \; ; \; v \in T_e G\}$.

To better understand all this, let us consider the next concrete situations, basic for our book.

(4.12) EXAMPLE

Let $H^d(n,p;K)$ be the vector space of all mappings $u: K^n \to K^p$ whose components $u_1, \ldots, u_p$ are homogeneous polynomials in $x_1, \ldots, x_n$ of degree d $(K = \mathbb{R}, \mathbb{C})$.

Consider first the analog of *L*-equivalence, namely the action

$$\lambda: G\ell(p,K) \times H^d(n,p;K) \to H^d(n,p;K)$$

$$(A,u) \longmapsto A \circ u.$$

It is well-known [Ad], p.12 that the Lie algebra $T_e G\ell(p,K)$ is the Lie algebra of all $p \times p$ matrices over K and that the exponential map is given by

(4.13) $$\exp A=1+\frac{A}{1!}+\frac{A^2}{2!}+\ \dots$$

It follows that

$$\frac{d}{dt}[(\exp(tA)\circ u]_{t=0}=A\circ u$$

for any polynomial map u. Hence

(4.14) $$T_u(G\ell(p,K)\cdot u)=\langle u\rangle e_1+\dots+\langle u\rangle e_p\subset H^d(n,p;K)$$

where $\langle u\rangle=K\langle u_1,\dots,u_p\rangle$ denotes the vector space spanned by the components u_i of u and $e_k=(0,\dots,0,1,0,\dots,0)$ with p entries and the k-th entry equal to 1 for $k=1,\dots,p$ form the canonical basis of K^p.

We shall also use the more condensed notation

(4.15) $$T_u(G\ell(p,K)\cdot u)=\langle u\rangle\cdot K^p$$

Let us consider a second action on $H^d(n,p;K)$, the analogue of $\mathcal{R}$ equivalence:

$$\rho: G\ell(n,K)\times H^d(n,p;K)\ \to\ H^d(n,p;K)$$

$$(A,u)\longmapsto u\circ A^{-1}$$

Using (4.13), it follows easily that

$$\frac{d}{dt}[\exp(tA)\cdot u(x)]_{t=0}=\frac{d}{dt}[u(\exp(-tA)\cdot x)]_{t=0}=$$

$$=\sum_{i,j=1,\dots,n}-a_{ij}x_j\,\frac{\partial u}{\partial x_i}(x)$$

Hence

(4.16) $$T_u(G\ell(n,K)\cdot u)=K\langle x_j\,\frac{\partial u}{\partial x_i};\ i,j=1,\dots,n\rangle$$

where one should think of $\frac{\partial u}{\partial x_i}$ as a column vector and of x_j as a scalar which multiplies each component of this vector.

We can combine these two actions λ and ρ to get a new action α, the analogue of $\mathcal{A}$-equivalence:

$\alpha: G\times H^d(n,p;K) \to H^d(n,p;K)$, where $G=G\ell(n,K)\times G\ell(p,K)$ and the rule is

$((A,B),u) \longmapsto B\circ u\circ A^{-1}$

It is a simple general fact that in such a case the tangent spaces also add, namely

(4.17) $\quad T_u(G\cdot u)=T_u(G\ell(n,K)\cdot u)+T_u(G\ell(p,K)\cdot u)=K<x_j \frac{\partial u}{\partial x_i}>+<u>K^p$.

Why is it useful to compute tangent spaces to orbits in general? A first information we get in this way is clearly the dimension of the orbits (for some concrete computations see next Chapter).

A second, more subtle information concerns the local structure of a *smooth* action $m: G\times M \to M$ around a given point $x\in M$. To explain this in detail, we start with the following.

(4.18) DEFINITION

A smooth submanifold $N\subset M$ is called a G-*transversal* (or G-*slice*) *at the point* x if $x\in N$ and $T_xN+T_x(G\cdot x)=T_xM$.

In other words N is a submanifold (usually regarded as a germ at x) which is transverse at x to the G-orbit of x. When the action m is clear from the context, we call N simply a transversal (slice) at the point x.

It is clear from the definition that dim N≥codim Gx and, in case of equality, we call N a *minimal* transversal (slice).

The isotropy group of x, namely $G_x=\{g\in G;\ g\cdot x=x\}$ is a Lie subgroup in G. Let $H\subset G$ be a submanifold such that $e\in H$ and $T_eH \oplus T_eG_x=T_eG$, where $\oplus$ denotes direct sum. In particular, note that

$\dim H = \dim G - \dim G_x = \dim(G\cdot x)$

To show the existence of such a submanifold H note that we can work locally around $e\in G$ in a coordinate neighbourhood.

Let S be a minimal slice at the point $x\in M$. The next result shows that a suitable *product structure*, compatible in a certain sense with the action m, occurs on a neighbourhood of x.

(4.19) PROPOSITION

Let $\bar{m}: H\times S \to M$ *be the map induced by the multiplication* m. *Then there are neighbourhoods* V_1 *of* e *in* H, V_2 *of* x *in* S *and* U *of* x *in* M *such that* $\bar{m}$ *maps* $V_1\times V_2$ *diffeomorphically onto* U.

PROOF

Note that $\dim(H\times S)=\dim M$ and hence, by Inverse Function Theorem, it is enough to show that the differential $d\bar{m}(e,x)$ is an epimorphism. The obvious relations

$$d\bar{m}(e,x)(T_eH\times 0)=T_x(G\cdot x)$$

$$d\bar{m}(e,x)(0\times T_xS)=T_xS$$

and the defining property of the slice S show that this is indeed the case.

(4.20) COROLLARY

With the above notations, any orbit which intersects U, *intersects also transversally the minimal slice* S.

This property will be used several times in the next Chapter (in relation with adjacency and specialization of orbits).

§3. TANGENT SPACES TO ORBITS IN JET SPACES

In this section we come back to the actions on jet spaces introduced at (3.20)-(3.23) and compute the tangent spaces to the corresponding orbits using the same ideas as in Example (4.12).

To do this, we first investigate the exponential map of the groups $M^k_{n,p}$ and D^k_n involved in these actions. For the matrix group $M^k_{n,p}$ the answer is quite simple. The Lie algebra $T_eM^k_{n,p}$ is nothing else but the Lie algebra of all $p\times p$ matrices with entries in E_n/m^k_n. The corresponding exponential map is given by the formula (4.13) as for usual matrix (i.e. linear) groups, with the necessary truncations.

If we consider the action of type λ

$$M^k_{n,p}\times J^k(n,p) \to J^k(n,p), \quad (j^{k-1}A, j^kf)\longmapsto j^k(A\cdot f)$$

it follows as in (4.15) that

(4.21) $$T_{j^kf}(M^k_{n,p}\cdot j^kf)=j^k(I_f\cdot E_{n,p})$$

where I_f is the ideal generated in E_n by the components $f_1,\ldots$ $\ldots,f_p$ of the map germ f and $I_f\cdot E_{n,p}$ denotes the E_n-submodule in $E_{n,p}$ generated by the products $a\cdot g$ with $a\in I_f$, $g\in E_{n,p}$.

For the group D^k_n of k-jets of diffeomorphisms the construction of the exponential map is more subtle. We have an obvious equality of vector spaces $T_eD^k_n=J^k(n,n)$.

An element $j^kv\in J^k(n,n)$ corresponds to a germ $v=\sum_{i=1,n} v_i\frac{\partial}{\partial x_i}$ of vector field at the origin of K^n with $v(0)=0$. To such a germ v corresponds in turn a 1-parameter family of germs of diffeomorphisms $g_t\in D_n$, where $g_t(x)$ $(t\in K)$ is the integral curve of the vector field v with the initial condition $g_o(x)=x$.

Indeed, it can be shown that for any $t\in K$ there is a small enough neighbourhood U of 0 in K^n on which g_t is defined (for more details, if necessary, one can consult [Mr], Chap. II). The map $t\longmapsto j^kg_t$ is then a 1-parameter subgroup in the group D^k_n and the corresponding tangent vector is precisely j^kv. In other words:

(4.22) $$\exp(t\cdot j^kv)=j^kg_t$$

We compute now the tangent spaces for the R_n^k-orbits using (4.22), in a similar way to the derivation of (4.16). Namely

$$\frac{d}{dt}j^k(f(\exp(-tj^k v(x))))_{t=0}=j^k(\frac{d}{dt}(f\circ g_{-t})(x)|_{t=0})=$$

$$=-j^k(df(x)(v(x)))=-j^k(\sum_{i=1,n} v_i(x)\frac{\partial f}{\partial x_i}(x)).$$

If J_f denotes the E_n-submodule in $E_{n,p}$ generated by the (column vectors) $\frac{\partial f}{\partial x_i}$ for i=1,...,n, then the above formula implies

(4.23) $\quad T_{j^kf}(R_n^k\cdot j^kf)=j^k(m_n\cdot J_f)$

We remark that a germ $f\in E_{n,p}^o$ induces a K-algebra morphism $f^*:E_p \to E_n$ via which any E_n-module acquires a structure (depending surely of f) of an E_p-module.

Using this remark and the formulas (4.21), (4.22) and (4.23) we leave to the reader the proof of the similar formulas

(4.24) $\quad T_{j^kf}(L_p^k\cdot j^kf)=j^k(m_p\cdot E_{n,p})$

$$T_{j^kf}(A_{n,p}^k\cdot j^kf)=j^k(m_n\cdot J_f+m_p\cdot E_{n,p})$$

$$T_{j^kf}(K_{n,p}^k\cdot j^kf)=j^k(m_n\cdot J_f+I_f\cdot E_{n,p})$$

In view of these equalities and (4.23) it is natural to introduce the following vector subspaces in $E_{n,p}^o$ for any map germ $f\in E_{n,p}^o$:

(4.25) $\quad T\mathcal{R}f=m_n\cdot J_f\ ,\quad T\mathcal{L}f=m_p\cdot E_{n,p}$

$$T\mathcal{A}f=m_n\cdot J_f+m_p\cdot E_{n,p}\ ,\quad T\mathcal{K}f=m_n\cdot J_f+I_f\cdot E_{n,p}\ .$$

More precisely, $T\mathcal{R}f$ and $T\mathcal{K}f$ are E_n-submodules in $E_{n,p}^o$, while $T\mathcal{L}f$ and $T\mathcal{A}f$ are only E_p-submodules. This is a basic difference which makes the first two equivalence relations much easier to study than the last two ones.

An obvious reason (but not the only one!) to consider these submodules is to have the formula

$$j^k(T\mathcal{R}f)=T_{j^kf}(R_n^k\cdot j^kf)$$

and the other three similar ones.

§4. MATHER'S LEMMA

Let $m:G\times M\to M$ be a smooth action. In order to decide whether two elements $x_o,x_1\in M$ are G-equivalent, the first idea is to try to find a path (a *homotopy*) $P=\{x_t;\ t\in[0,1]\}$ such that P is entirely contained in a G-orbit.

It turns out that this naive approach works quite well and the next result gives effective necessary and sufficient conditions for a connected submanifold (in our case the path P) to be contained in an orbit.

(4.26) MATHER'S LEMMA [Ma]

Let $m:G\times M\to M$ *be a smooth action and* $P\subset M$ *a connected smooth submanifold. Then* P *is contained in a single* G*-orbit if and only if the following conditions are fulfilled:*

(a) $T_x(G\cdot x)\supset T_xP$, *for any* $x\in P$.

(b) $\dim T_x(G\cdot x)$ *is constant for* $x\in P$.

PROOF

The necessity of the conditions (a) and (b) is obvious.

To prove the sufficiency, recall first that the map $m_x:G\to M$, $g\longmapsto g\cdot x$ has the property (4.10). Using this, the conditions (a) and (b) above become

(a') $dm_x(e)(T_eG)\supset T_xP$, for $x\in P$

(b') $\operatorname{rank}(dm_x(e))=\text{constant}$, for $x\in P$.

We fix a scalar product on the tangent space T_eG and let

L_x denote the orthogonal complement to $\ker(dm_x(e))$ for $x \in P$. Then $L = \bigcup_{x \in P} (x \times L_x)$ is the total space of a sub vector bundle of the trivial bundle $P \times T_eG \to P$. If we define

$$L_x^o = L_x \cap (dm_x(e))^{-1}(T_xP) \text{ and } L^o = \bigcup_{x \in P} (x \times L_x^o)$$

then L^o is a sub vector bundle in L and the differentials $dm_x(e)$ induce an isomorphism of vector bundles over P, namely $L^o \xrightarrow{\simeq} TP$. Let $f: TP \to L^o$ be the inverse of this isomorphism and $p: P \times T_eG \to T_eG$ be the second projection.

Note that $p \circ f$ is a smooth map satisfying

$$dm_x(e)(p \circ f(v)) = v, \qquad \text{for } v \in T_xP.$$

The manifold P being connected, any two points of it can be joined by a smooth path in P. Hence it is sufficient to show that any smooth path $c:[0,1] \to P$ has the image contained in a single G-orbit.

Let $\dot{c}(t) \in T_{c(t)}P$ be the corresponding tangent vector and put $X_t = pf(\dot{c}(t)) \in T_eG$.

Let $\bar{X}_t$ be the right invariant vector field on the Lie group G extending the vector X_t. In other words $\bar{X}_t(e) = X_t$ and $\bar{X}_t(g) = dR_g(e)(X_t)$, where $R_g(h) = gh$.

For $t_o \in [0,1]$, consider the ordinary differential equation $\dot{a}(t) = \bar{X}_t(a(t))$, $a(t_o) = e$. For $\varepsilon > 0$ sufficiently small, there is a solution $a:(t_o - \varepsilon,\ t_o + \varepsilon) \to G$ of this equation. A straightforward computation shows that

$$\frac{d}{dt}(a(t)^{-1} \cdot c(t)) = 0$$

and hence $a(t)^{-1} \cdot c(t) = c(t_o)$. Therefore $c(t)$ stays in a single orbit for $|t - t_o| < \varepsilon$. Since the interval $[0,1]$ is connected, this implies that the image of c is indeed contained in a G-orbit.

(4.27) EXERCISE

Let $M = \mathbb{R}^2$,

$G=\{\begin{pmatrix} a & b \\ 0 & c \end{pmatrix};\ ac\neq 0\}$ and $m: G\times M \to M$

be the usual linear action induced by $G\subset G\ell(2,\mathbb{R})$.

(i) Show that there are exactly 3 orbits: the origin, the x-axis minus the origin and M minus the x-axis.

(ii) Show that $P=\{(x,y)\in M;\ y-(x-1)^2=0\}$ is a connected submanifold in M which satisfies the condition (a) in (4.26) but is not contained in a single orbit.

(iii) Give a concrete example of a smooth action $G\times M \to M$ and of a connected submanifold $P\subset M$ which satisfies the condition (b) in (4.26) but is not contained in a single orbit.

Therefore one needs both conditions (a) and (b) in Mather's Lemma.

(4.28) REMARK

A statement slightly more general than (4.26) and one of its applications can be found in [D3].

In the applications of Mather's Lemma to be given in this book, one deals with the following situation.

A Lie group G acts smoothly on the manifolds M and N and $p: M \to N$ is an *equivariant* submersion, i.e. p is a submersion at any point $x\in M$ and $p(g\cdot x)=g\cdot p(x)$ for any $g\in G$, $x\in M$.

(4.29) COROLLARY

With the above notations, for any point $y\in N$ *and any connected open subset* P *in* $p^{-1}(y)$ *one has the equivalence:*

The submanifold P *is contained in a single* G*-orbit if and only if the condition* (a) *in* (4.26) *holds.*

PROOF

First note that P (if not empty) is indeed a submanifold in M. Next, using condition (a) we get

$$\dim T_x(G\cdot x)=\dim T_xP+\dim T_y(G\cdot y)$$

which clearly implies the condition (b) in (4.26).

CHAPTER 5

BASIC CLASSIFICATION EXAMPLES IN LINEAR ALGEBRA AND ALGEBRAIC GEOMETRY

§1. LINEAR MAPS AND QUADRATIC FORMS

In this Chapter we specialize the actions introduced in Example (4.12) to some important concrete cases, in which it is possible to give a *complete list of the orbits*. Besides this listing, we introduce an *hierarchy among these orbits* (equivalence classes) which, roughly speaking, describes the possible changes in these classes under small deformations.

(5.1) DEFINITION

Let $G\times M \to M$ be an algebraic action. We say that *an orbit* $G\cdot x$ (resp. *the element, normal form* x) *specializes to an orbit* $G\cdot y$ (resp. *to the element, normal form* y) if $\overline{G\cdot x}\supset G\cdot y$. This situation is also described by saying that *the orbit* $G\cdot y$ (resp. *the element* y) *is adjacent to the orbit* $G\cdot x$ (resp. *to the element* x). One assumes here $Gx\neq Gy$.

Notation: $x \to y$, $Gx \to Gy$.

Since the orbits are semialgebraic sets (4.8), it is clear by the *Curve Selection Lemma* [M1], p.25 that $x \to y$ if and only if there is a real analytic path $c:[0,\varepsilon) \to M$ for some $\varepsilon>0$ such that $y=c(0)$ and $c((0,\varepsilon))\subset G\cdot x$. Hence y can be regarded as a limit of points $x_t=c(t)\in G\cdot x$ for $t \to 0$.

A necessary condition for $x \to y$ is obviously $\dim G\cdot x > \dim G\cdot y$ by our Remark (4.9).

The first example we treat is *the classification of linear maps*, i.e. $G=G\ell(n,K)\times G\ell(p,K)$, $M=\mathrm{Hom}(K^n,K^p)=H^1(n,p;K)$ and the action is that introduced in (3.3.a) or in (4.12) as action α for d=1.

Any linear map $u\in M$ is equivalent to a normal form

$u_k(x_1,\dots,x_n)=(x_1,\dots,x_k,0,\dots,0)$ where $k=\text{rank } u\le m=\min(n,p)$. Using (4.17) it follows that

$$T_{u_k}(Gu_k)=K<x_je_i;\ i=1,\dots,k;\ j=1,\dots,n>+$$
$$+<x_1,\dots,x_k>K^p$$

Hence a minimal slice S at the point u_k can be given by

(5.2) $\qquad S=\{u_k+\sum a_{ij}x_ie_j;\ k<i\le n,\ k<j\le p \text{ and } a_{ij}\in K\}$

Note that dim $S=(n-k)(p-k)$ and hence $G\cdot u_k$ is a smooth manifold of dimension np-dim S. As Gu_k is precisely the set of linear maps in M of rank equal to k, we have thus obtained a proof of a well-known exercise (e.g. [GG], p. 12).

We have also the next result

(5.3) PROPOSITION

The specialization relations among the normal forms u_k *are the following*

$$u_m \to u_{m-1} \to \dots \to u_1 \to u_o=\{0\}$$

In other words: $\overline{Gu_k}=Gu_k\cup Gu_{k-1}\cup\dots\cup Gu_1\cup\{0\}$. *Moreover,* $\overline{Gu_m}=M$ *is smooth and the singular set* $S(\overline{Gu_k})$ *of the closure* $\overline{Gu_k}$ *for* $k<m$ *is precisely* $\overline{Gu_{k-1}}$.

PROOF

From (5.2) we clearly have $v\in S \Longrightarrow \text{rank } v\ge \text{rank } u_k$. Conversely, for $k<s\le m$ the element

$$u_s(t)=u_k+tx_{k+1}e_{k+1}+\dots+tx_se_s$$

has rank s and $\lim_{t\to 0} u_s(t)=u_k$.

This proves the first part of the statement as well as the equality $\overline{Gu_m}=M$.

To determine $S(\overline{Gu_k})$ we proceed as follows. First Gu_k is a smooth semialgebraic set and an open subset in $\overline{Gu_k}$ and hence

$$S(\overline{Gu_k}) \subset \overline{Gu_k} \setminus Gu_k$$

Since the singular set is closed, to prove the converse inclusion it is enough to show that $S(\overline{Gu_k}) \supset Gu_{k-1}$ (note the implicit use of the specialization relations!).

We do this in detail in the special case k+1=n=p and leave the interested reader to complete the proof in the general case along the same lines.

Hence from now on $D=\overline{Gu_{n-1}}$ is exactly the hypersurface det A=0 of all degenerate square matrices A in $Hom(K^n,K^n)$. Then the slice S from (5.2) at the point u_{n-2} is formed by matrices

$$u(a,b,c,d) = \begin{pmatrix} I_{n-2} & & 0 \\ & & \\ & a & b \\ 0 & c & d \end{pmatrix}$$

with $a,b,c,d \in K$.

It is clear that

$$u(a,b,c,d) \in Gu_{n-1} \iff (a,b,c,d) \neq 0 \quad \text{and } ad-bc=0$$

$$u(a,b,c,d) \in Gu_{n-2} \iff (a,b,c,d) = 0$$

By (4.19) it follows that, locally around u_{n-2}, the set $\overline{Gu_{n-1}}$ is the product of the 3-dimensional singular affine quadric $Q: ad-bc=0$ in K^4 with a smooth (n^2-4)-dimensional space V.

Since Q is singular at the origin (Exercise!), it follows that $Q \times V$ is singular at any point of $0 \times V$ and hence $S(\overline{Gu_{n-1}}) \supset$ $\supset Gu_{n-2}$.

In particular, it follows that

$$(5.4) \qquad \operatorname{codim}_D S(D) = \dim D - \dim S(D) = 3$$

and hence, in the complex case, D is a *normal hypersurface* see [KK] p.315 or Proposition (10.3) in our book.

Before ending the discussion of this example, we notice that in the *complex case* all the orbits Gu_k are *connected* (since the linear groups $G\ell(n,\mathbb{C})$ are connected!).

In the *real case* it is well-known that the open orbit $\mathrm{Hom}(\mathbb{R}^n, \mathbb{R}^n) \setminus D$ has exactly *two connected components*, corresponding to the inequalities $\det A>0$ and $\det A<0$.

We turn now briefly to the *classification of quadratic forms*. More precisely, we consider the action ρ from Example (4.12) in the case $p=1$, $d=2$. By *Sylvester Theorem* (see for instance [L], p.365) concerning the normal forms of *real* quadratic forms, it follows that the orbits in this case are exactly

(5.5) $\quad U_{i,r}=G\ell(n, \mathbb{R})\cdot f_{i,r}$, where $0\leq i\leq r\leq n$

and $f_{i,r}(x)=-x_1^2-\ldots-x_i^2+x_{i+1}^2+\ldots+x_r^2$.

The positive integers i and r are called the *index* and respectively the *rank* of the quadratic form $f_{i,r}$ and they give in fact a *complete set of invariants* for this action (i.e. two real quadratic forms are in the same orbit if and only if they have the same index and rank).

Using the formula (4.16) we get

$$T_{f_{i,r}}U_{i,r}=\mathbb{R}\langle x_a x_b\rangle, \quad \text{with } 1\leq a\leq n, \quad 1\leq b\leq r.$$

In particular $\dim U_{i,r}=\frac{r(2n-r+1)}{2}$ and hence $U_{i,r}$ is an open orbit if and only if $r=n$.

(5.6) EXERCISE

(i) Let G_0 denote the connected component of the identity element in the group $G\ell(n, \mathbb{R})$.

Show that $G_0\cdot f=G\ell(n, \mathbb{R})\cdot f$ for any quadratic form f. Deduce that the orbits $U_{i,r}$ are all connected.

(ii) Determine all the specialization relations among the normal forms $f_{i,r}$. Hint: show that $f_{i,r}\to f_{j,s}$ if and only if $i\geq j$, $r\geq s$ and $r-i\geq s-j$.

Note that in the *complex case* a complete set of invariants for quadratic forms is provided just by the rank and hence there are only n+1 orbits.

(5.7) EXERCISE

State and prove the analogue of (5.3) in the case of classification of complex quadratic forms.

One of the conclusions of this section should be that the classification problems in the real case are usually more complicated than in the complex case: some $\pm$ signs occur and the orbits are not necessarily connected.

§2. CUBIC FORMS AND PLANE CUBIC CURVES

In this section we consider in detail the action ρ from (4.12) in the special case $K=\mathbb{C}$, $d=3$, $p=1$ i.e. classification of complex cubic forms in n variables. This example will bring some new features compared to those in the previous section, namely:

(i) *occurrence of infinitely many orbits;*

(ii) *substantial connections with projective Algebraic Geometry*, and

(iii) *subtlety of specializations.*

We show first that for $n\geq 3$, this action has infinitely many orbits. To show this, it is enough to prove the nonexistence of open orbits for $n\geq 3$. Indeed, since the orbits are semialgebraic subsets in $H^3=H^3(n,1;\mathbb{C})$, it follows then that they are not even countable i.e. some continuous parameters (called *moduli*) occur in describing these orbits.

It is well known that

$$\dim G\ell(n,\mathbb{C})=n^2, \quad \dim H^3=\frac{n(n+1)(n+2)}{6}$$

The orbit corresponding to the cubic form $f\in H^3$ is open if and only if the differential $d\rho_f(e)$ is surjective, where $\rho_f:G\ell(n,\mathbb{C})\to H^3$, $\rho_f(g)=f\circ g^{-1}$ (4.10). This implies the inequality

$$n^2\geq\frac{n(n+1)(n+2)}{6}$$

and this holds only for $n=1,2$. In these special cases, the orbits correspond to the following normal forms (in particular, there are indeed open orbits in these two cases).

(5.8) TABLE

n	Normal form	Dimension of the orbit
1	x^3	1
	0	0
2	x^2y+y^3	4
	x^2y	3
	x^3	2
	0	0

The proof is an easy exercise for the reader (if necessary, have a look at [Gi], p.66 where the slightly more complicated case of real cubic forms is treated).

We concentrate now on the first difficult case $n=3$ in order to understand the connection between this type of actions and basic classification problems in Algebraic Geometry.

Let $\mathbb{P}^m$ be the *complex projective m-dimensional space.* A homogeneous polynomial $f \in H^d(n,1;\mathbb{C})$ with $n=m+1$ defines a *projective hypersurface*

(5.9) $$V=V(f)=\{x\in \mathbb{P}^m;\ f(x)=0\}$$

More precisely, we regard V as a subvariety in $\mathbb{P}^m$ corresponding to the principal ideal $I_V=(f)$ in $\mathbb{C}[x_1,\dots,x_n]$. The degree $d=\deg(f)$ is also called the *degree of the hypersurface* V. Note also that V is not necessarily reduced.

It is known that any automorphism of the projective space $\mathbb{P}^m$ is induced by a linear map in $G\ell(n,\mathbb{C})$, [Ha], p.151. This remark makes natural the following.

(5.10) DEFINITION

Two subvarieties $A,B\subset \mathbb{P}^m$ defined respectively by the ideals I_A and I_B in $\mathbb{C}[x_1,\dots,x_n]$ are *projectively equivalent* if there is a linear map $g \in G\ell(n,\mathbb{C})$ such that $g^*(I_A)=I_B$. Notation: $A \sim B$.

Many problems in Algebraic Geometry consider a given class C of subvarieties in $\mathbb{P}^m$ (e.g. hypersurfaces of degree d) and ask for a list of equivalence classes in $C/\sim$.

One has the following easy

(5.11) LEMMA

For two homogeneous polynomials $f,g \in H^d(n,1;\mathbb{C})$ *the following are equivalent*

(i) f *and* g *are* ρ-*equivalent as in* (4.12).

(ii) f *and* g *are* α-*equivalent as in* (4.12).

(iii) $V(f) \sim V(g)$.

PROOF

Exercise for the reader.

In order to solve the *global* problem of classifying hypersurfaces modulo $\sim$, it is necessary first to consider some *local* (and even *punctual*) properties for hypersurfaces.

A point $x \in V(f)$ is *singular* if $df(x)=0$ (working in an affine coordinate neighbourhood of x one *should check* that this condition is equivalent to the germ of analytic space $(V(f),x)$ being singular as in (3.14)).

The *hessian* of the polynomial f is the polynomial

$$(5.12) \qquad H(f)=\det\left|\frac{\partial^2 f}{\partial x_i \partial x_j}\right|_{1\leq i,j\leq n}$$

which is homogeneous of degree $n(d-2)$.

A point $x \in V(f)$ is an *inflexion point* if x is not singular and $H(f)(x)=0$.

We restrict from now to the case of plane curves (m=2). Then an inflexion point x is exactly a simple point $x \in V(f)$ with the *order of contact* between the curve $V(f)$ and the tangent line $T_xV(f)$ at least 3 (for notions and results on plane curves we refer to the beautiful recent book [BK]).

Assume from now on also that d=3, hence $V=V(f)$ is a

(plane) cubic curve. Then deg $H(f)=3$ and by *Bézout Theorem* we expect a *general* cubic curve V to have 9 inflexion points. It turns out thatthis is true for any smooth cubic curve V and that the nine inflexion points of V have the following property: any line in $\mathbb{P}^2$ determined by two of them contains exactly a third inflexion point on it.

A simple lemma shows that for any set of nine points in $\mathbb{P}^2$ with this property, there is a linear coordinate system on $\mathbb{P}^2$ such that these 9 points have the following coordinates

$$(5.13)\qquad (0:1:\varepsilon_j),\quad (\varepsilon_j:0:1),\quad (1:\varepsilon_j:0)$$

where $j=1,2,3$ and ε_j are the cubic roots of -1. Any cubic curve in $\mathbb{P}^2$ passing through these nine points is given by an equation (*Hesse normal form* [BK], p.293).

$$(5.14)\qquad C_t: x^3+y^3+z^3+3txyz=0$$

with $t\in\mathbb{C}\cup\{\infty\}$ (which is the cubic curve corresponding to $t=\infty$?)

And C_t is smooth if and only if $t\in\mathbb{C}$ and $t^3\neq-1$.

In other words, we have shown that any smooth cubic curve is projectively equivalent to the cubic in normal form C_t. Moreover, it is known that two such curves C_t and C_s are equivalent if and only if $j(t)=j(s)$, where the rational function j (called the *j-invariant*) is given (see [BK], p.302) by the formula

$$(5.15)\qquad j(a)=\frac{a^3(8-a^3)^3}{(1+a^3)^3}$$

The complex number t is also called the *modulus* of the family C_t.

Let us consider now the singular cubic curves. First assume that C is an irreducible singular cubic curve. Then C has exactly one singular point which can be

(i) a *node* and then C is called *nodal* and is equivalent to the normal form $x^3+y^3-xyz=0$.

(ii) a *cusp* and then C is called *cuspidal* and is equivalent to the normal form $zy^2-x^3=0$. (If you do not know what a node or cusp is, you can have a look at Example (10.14) further in this book).

When C is reducible, its components must be either conics or lines and all the possibilities are described in the next table. The dimensions given here are the dimensions of the corresponding orbits. (If we add the trivial orbit 0, we get the complete list of these orbits in $H^3(3,1;\mathbb{C})$).

(5.16) TABLE

	Type	Normal form	Dimension	Picture
1.	smooth	$x^3+y^3+z^3+txyz$ $t\in\mathbb{C},\ t^3\neq-1$	9	
2.	nodal	x^3+y^3-xyz	9	
3.	cuspidal	zy^2-x^3	8	
4.	conic+chord	x^3+xyz	8	
5.	conic+tangent	x^2y+y^2z	7	
6.	triangle	xyz	7	
7.	3 concurrent lines	x^3+y^3	6	
8.	2 lines, one of them double	x^2y	5	
9.	a triple line	x^3	3	

Though the listing contained in Table (5.16) is well--known, the next result seems to be new.

(5.17) PROPOSITION

The specializations among plane cubic curves are described by the following diagram

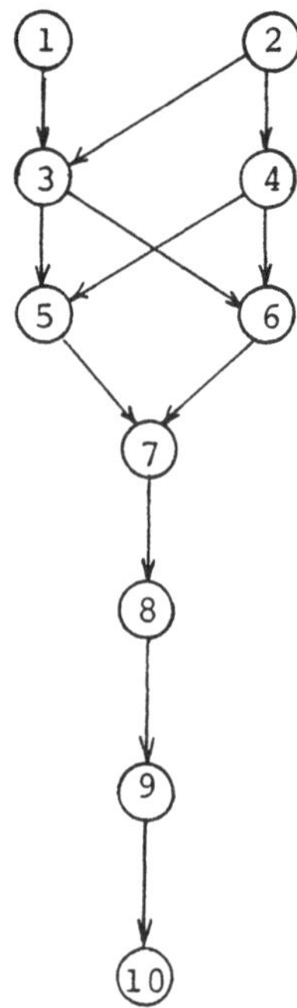

Here (k) denotes the orbit in $H^3(3,1;\mathbb{C})$ corresponding the k-th normal form in Table (5.16). By (1) it is meant the orbit corresponding to a *fixed* $t_o \in \mathbb{C}$, $t_o^3 \neq -1$ and (10) stands for the trivial orbit 0.

Before giving the proof, we note that *some specializations can be represented intuitively very well by a picture.* And in this sense, the word "specialization" was already familiar to classical algebraic geometers. As an example consider the specializations

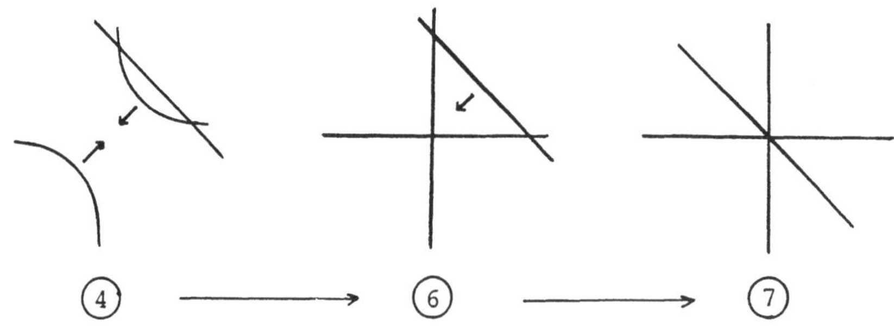

FIGURE (5.18)

Here the first specialization is given by deforming a smooth conic into a pair of lines (in the picture the hyperbola $xy=t$ goes into the coordinate axes $xy=0$ when $t \to 0$). Secondly, the chord is shifted so that the 3 lines become concurrent.

PROOF OF (5.17)

We give a detailed proof for ③ ⟶ ⑤ and ① ↛ ④ . The rest of the specializations can safely be left to the reader to check by a similar method.

Let W be the orbit ⑤ and $f=x^2y+y^2z$ the corresponding normal form. Then by (4.16) we get

$$T_fW=\mathbb{C}\langle x^3,y^3,x^2y,xy^2,y^2z,x^2z+2yz^2,xyz\rangle$$

Hence the general element in a minimal slice at the point f can be taken to be

$$F=f+az^3+bz^2y+cz^2x\ , \quad \text{with}\quad a,b,c\in\mathbb{C}.$$

To show that ③ ⟶ ⑤ it is necessary and sufficient to prove that, for suitable a,b,c as small as we want, the equation $F=0$ defines a cuspidal cubic curve.

First note that on the line $z=0$ there is no singular point for $V(F)$. Indeed

$$\frac{\partial F}{\partial y}(x,y,0)=\frac{\partial F}{\partial z}(x,y,0)=0 \implies x=y=0.$$

In the affine open subset $U\subset\mathbb{P}^2$ defined by $z\neq 0$ the curve is given by the equation

$$\bar{F}=F(x,y,1)=x^2y+y^2+a+by+cx=0\ .$$

Hence a singular point $(x_o:y_o:1)\in V(F)$ must satisfy the system

$$S:\begin{cases}\bar{F}=0\\ \bar{F}_x=2xy+c=0\\ \bar{F}_y=x^2+2y+b=0\end{cases}$$

In the new coordinates $X=x-x_o$, $Y=y-y_o$ we get by a simple computation

$$\bar{F}=y_o X^2+2x_o XY+Y^2+X^2Y$$

The singular point (located now at X=Y=0) is a cusp if $y_o=x_o^2$ and $x_o\neq 0$. This equation and the system S give

$$x_o=\sqrt{-\frac{b}{3}},\ y_o=-\frac{b}{3},\quad c=\frac{2b}{3}\sqrt{-\frac{b}{3}},\quad a=\frac{b^2}{3}$$

Hence, for this choice of the coefficients a and c as functions of b, the cubic V(F) is cuspidal. Since $b \to 0$ implies $a(b) \to 0$, $c(b) \to 0$ this ends the proof of (3) → (5) .

Now we show that (1) *does not specialize to* (4). As above, one shows that a slice at the normal form $g=x^3+xyz$ correspon-ding to the orbit (4) is determined by the cubic forms

$$G=g+ay^3+bz^3 \quad \text{for} \quad a,b\in\mathbb{C}.$$

For a=b=0, the cubic curve V(G)=V(g) is in the orbit (4). For (a=0 and b≠0) or for (a≠0 and b=0) the cubic V(G) is obviously nodal.

Assume now that ab≠0. Then V(G) is projectively equivalent to the cubic C_t from (5.14) with $t^{-1}=3\sqrt[3]{ab}$ (just use the coordinate change $X=x$, $Y=\sqrt[3]{a}\,y$, $Z=\sqrt[3]{b}\,z$). Then using the formula (5.15) for j(t), it follows that $j(t) \to \infty$ when $a,b \to 0$. Hence it is not possible to find a,b as small as we want and such that $j(t)=j(t_o)$ which is necessary for C_t to be a smooth cubic in the orbit (1).

We hope that this example is enough to convince the reader that the problem of the determination of all the specializations in a concrete case might be one of the most subtle questions.

§3. PENCILS OF QUADRICS

In this section we relate the action α from (4.12) on $H^d(n,p;\mathbb{C})$ for $p\geq 2$ with some classical objects in Algebraic Geometry, namely the *linear systems*.

For an element $f=(f_1,\ldots,f_p)\in H^d(n,p;\mathbb{C})$ we introduce the *ideal* I_f generated in $\mathbb{C}[x_1,\ldots,x_n]$ by the components $f_1,\ldots,f_p$ of f and the *algebraic subvariety* $V_f\subset\mathbb{P}^m$ corresponding to the ideal I_f (here again $m=n-1$).

(5.19) EXERCISE

Two elements $f,g\in H^d(n,p;\mathbb{C})$ are α-equivalent if and only if V_f and V_g are projectively equivalent. Hint: recall the proof of (3.16).

If $\dim\mathbb{C}<f_1,\ldots,f_p>=p\geq 2$, then in classical Algebraic Geometry one considers the *linear system* $L_f: a_1f_1+\ldots+a_pf_p=0$ of hypersurfaces in $\mathbb{P}^m$ parametrized by points $a=(a_1:\ldots:a_p)\in\mathbb{P}^{p-1}$. Then $(p-1)$ is called the *dimension* of the linear system L_f.

Two linear systems L_f and L_g are *equivalent*, by definition, just when f and g belong to the same α-orbit in $H^d(n,p;\mathbb{C})$.

The simplest linear systems are the 1-dimensional ones, called *pencils*. Some readers have maybe already noticed that we met a pencil in the formula (5.14) above!

We describe now briefly the classification of *pencils of quadrics*. This corresponds to the α-action on $H^2=H^2(n,2;\mathbb{C})$. Indeed an element $f=(f_1,f_2)\in H^2$ is just a pair of quadratic forms and if $\dim\mathbb{C}<f_1,f_2>\leq 1$ it is easy to see that f is α-equivalent to an element $(f_1,0)$, with the quadratic form $f_1=x_1^2+\ldots+x_r^2$ where $r=\text{rank}\,(f_1)$.

Assume from now on that $\dim\mathbb{C}<f_1,f_2>=2$, hence $L_f: a_1f_1+a_2f_2=0$ is indeed a pencil of quadrics.

An *algebraic derivation* of the normal forms for pencils of quadrics was already known from the last century and is not at all easy, see for instance [HP], Chap. XIII.

Without entering into the details of this classification, we want to explain the *geometric meaning* of it. For complete proofs, the reader can consult [D1].

To the pencil of quadrics L_f above correspond *two geometric objects*: the line L_f in the projective space $\mathbb{P}(H)$ (associated to the vector space $H=H^2(n,1;\mathbb{C})$) determined by the two points $f_1,f_2\in\mathbb{P}(H)$ and the algebraic subvariety $V_f\subset\mathbb{P}^m$ introduced above.

Recall that in H we have the orbits U_r corresponding to quadratic forms of rank r (see the end of section one in this Chapter). Let $W_r\subset\mathbb{P}(H)$ be the quasi-projective varieties corresponding to U_r, for $r=1,\ldots,n$.

Then the equivalence class of a pencil of quadrics L_f is completely determined by the position of the line L_f with respect to the varieties W_r and by the singular set $S(V_f)$ of the variety V_f. As an example, a *generic pencil* L_f corresponds to a line L_f which intersects W_{n-1} in exactly n points and $L_f\cap W_r=\emptyset$ for $r<n-1$. Moreover, in this case $S(V_f)=\emptyset$, i.e. V_f is a smooth complete intersection of two quadrics.

A normal form for this f is given by the formula

$$(5.20)\qquad f_1=x_1^2+\ldots+x_n^2\ ,\quad f_2=a_1x_1^2+\ldots+a_nx_n^2\ \text{ with } a_i\in\mathbb{C}$$
$$\text{and } a_i\neq a_j \quad\text{for}\quad i\neq j.$$

A detailed study of this type of singularity can be found in [Kn], where the singularity f is denoted by $\widetilde{D}_{n+1}$.

Let us consider now the particular case n=2. Then dim H=3 and hence $\mathbb{P}(H)$ is the 2-dimensional projective space, i.e. a plane. In this plane we have the conic

$$W_1: ac-b^2=0 \quad\text{where}\quad q=ax^2+2bxy+cy^2$$

is the equation for an element in H, i.e. *a binary quadratic form*.

(5.21) EXERCISE

Show that the following is a complete list of normal forms for pencils of binary quadratic forms: $(0,0)$, $(x^2,0)$,

$(xy,0)$, (x^2,xy), (x^2,y^2). Hint: The first three of these normal forms correspond to the classification of binary quadratic forms. The fourth case corresponds to a line L_f *tangent* to the conic W_1. And the last case corresponds to L_f being a *secant* (*chord*) for the conic W_1. If you need more details, look in [Gi] p.71.

§4. PENCILS OF BINARY CUBIC FORMS

The classification of pencils of binary cubic forms which we treat now has a nice geometric description (similar to that for pencils of quadrics) and is basic for the K-classification of the complex map germs in $E^o_{2,2}$, see [DG2], [DG3].

This classification problem corresponds by (5.19) to the listing of the α-orbits in $H^3(2,2;\mathbb{C})=H^3$. Here is the result.

(5.22) TABLE

	Normal form	Dimension
1.	$f_t=(x^3+3x^2y,\ 3xy^2+ty^3)$ $t\in\mathbb{C}\setminus\{0,1,9\}$	7
2.	$(x^3+x^2y,\ xy^2)$	7
3.	(x^3,y^3)	6
4.	(x^3,xy^2+y^3)	6
5.	(x^2y,xy^2)	6
6.	(x^3,xy^2)	6
7.	(x^3,x^2y)	5
8.	$(x^2y+y^3,0)$	5
9.	$(x^2y,0)$	4
10.	$(x^3,0)$	3
11.	$(0,0)$	0

The dimensions given here are the dimensions of the corresponding α-orbits in H^3. Since $\dim H^3=8$, there is no open orbit.

Two normal forms f_t and f_s above are in the same α-orbit if and only if $j(t)=j(s)$ where

$$j(t)=\frac{t^2(t-9)}{t-1}$$

Details can be found in [DG3].

The method for getting the normal forms in Table (5.22) is in principle the same as when classifying pencils of binary quadratic forms (5.21), but more subtle. That is why we give now all the details.

PROOF

To a general binary cubic form

$$C=ax^3+3bx^2y+3cxy^2+dy^3$$

we associate its *hessian* (compare to (5.12)) $H(C)=36(\alpha x^2+\beta xy+\gamma y^2)$, with $\alpha=ac-b^2$, $\beta=ad-bc$, $\gamma=bd-c^2$ and its discriminant $\Delta=\beta^2-4\alpha\gamma$.

In the complex projective space $\mathbb{P}^3=\mathbb{P}(H^3(2,1;\mathbb{C}))$ of all nonzero binary cubic forms, the form C above corresponds to the point $(a:b:c:d)$.

In these coordinates on $\mathbb{P}^3$ we consider the *quartic surface* $V:\Delta=0$ and the *twisted cubic curve* $W:\alpha=\beta=\gamma=0$ (for this curve see [Ha], p.12-14)

The natural action $G\times\mathbb{P}^3\to\mathbb{P}^3$ with $G=G\ell(2,\mathbb{C})$ coming from the corresponding ρ-action on $H^3(2,1;\mathbb{C})$ has exactly 3 orbits by (5.8):

$$G\cdot x^3=W\ ,\quad G\cdot x^2y=V\setminus W\ ,\quad G\cdot(x^2y+y^3)=\mathbb{P}^3\setminus V\ .$$

It follows easily that W and V are G-invariant subvarieties in $\mathbb{P}^3$ and that W is precisely the singular part S(V) of the surface V (Exercise).

If $f=(f_1,f_2)\in H^3$ and $\dim\mathbb{C}\langle f_1,f_2\rangle\le 1$, then f is α-equivalent to a normal form $(f_1,0)$ and this case gives the last 4 types in Table (5.22).

We assume from now on that $\dim \mathbb{C}<f_1,f_2>=2$ i.e. f_1 and f_2 determine distinct points in $\mathbb{P}^3$. Let L_f denote the line determined by these points $[f_1]$, $[f_2] \in \mathbb{P}^3$. It is clear that the position of the line L_f with respect to the varieties V, W is an invariant of the α-class of f and it will turn out that *this position (suitably interpreted) completely determines this class.*

The multiplicities of the points of intersection in $L_f \cap V$ are described by the multiplicities of the zeros of the binary form $\Delta_f(u,v)=\Delta(uf_1+vf_2)$ of degree 4.

We need the following.

(5.23) LEMMA

The intersection $L_f \cap V$ *contains at least two distinct points.*

PROOF

Let us suppose that $L_f \cap V$ consists of exactly one point f_1, which should exist by general reasons ([Ha], p.48). Using the homogeneity of $\mathbb{P}^3$ with respect to the above G-action we can assume that either $f_1=x^2y$ or $f_1=x^3$.

We give the proof only in the first case, the second one being completely similar. We work in the affine open set U in $\mathbb{P}^3$ defined by $b \neq 0$.

Under the usual identification $U=\mathbb{C}^3$, f_1 corresponds to the origin. We can obviously assume that the second defining point f_2 of L_f belongs to U. Hence, in the affine coordinates: $f_2=(a,c,d)$.

Next we have:

$$L_f \cap U=\{t(a,c,d);\ t \in \mathbb{C}\}$$

$$V \cap U=\{(a,c,d);\ \Delta(a,1,c,d)=0\}$$

Hence the intersection $L_f \cap V \cap U$ is described by the following equation in t: $(a^2d^2+4ac^3)t^4-6acdt^3-3c^2t^2+4dt=0$.

Since this equation should have only the solution t=0, it follows that all but one of its coefficients must vanish. If $d \neq 0$, then c=a=0 and $L_f=(x^2y,y^3)$ clearly intersects V in two points, namely corresponding to x^2y and y^3. If $c \neq 0$, then d=a=0 and $L_f=(x^2y,xy^2)$ meets again V in two points.

In the remaining case c=d=0, we have $L_f \subset V$. So we reach a contradiction, and this ends the proof of our Lemma.

Using this result, we assume that the element $f=(f_1,f_2)\in H^3$ satisfies $f_1,f_2\in V$. Moreover, let D be the *effective divisor* of degree 4 induced by the surface V on the line L_f. To determine this divisor, one has to compute for each normal form f the corresponding binary form Δ_f.

Using this information, we can list the orbits as follows (*the points* $P_i\in L_f\cap V$ *are distinct and in* $V\setminus W$ *if not stated otherwise*).

(a) D *is not defined* $\Longleftrightarrow$ $f\sim(x^3,x^2y)$.
($L_f\subset V$)

(b) $D=P_1+P_2+P_3+P_4$ $\Longleftrightarrow$ $f\sim f_t$, $t\in\mathbb{C}\setminus\{0,1,9\}$.

(c) $D=2P_1+P_2+P_3$ $\Longleftrightarrow$ $f\sim(x^3+x^2y,\ xy^2)$

(c') $D=2P_1+P_2+P_3$, $P_1\in W$ $\Longleftrightarrow$ $f\sim(x^3,xy^2+y^3)$

(d) $D=2P_1+2P_2$ $\Longleftrightarrow$ $f\sim(x^2y,xy^2)$

(d') $D=2P_1+2P_2$, $P_1,P_2\in W$ $\Longleftrightarrow$ $f\sim(x^3,y^3)$

(e) $D=3P_1+P_2$, $P_1\in W$ $\Longleftrightarrow$ $f\sim(x^3,xy^2)$

These cases (except for case (a)!) can be represented by the next self-explanatory pictures (5.24).

An unexpected consequence of these results for the geometry of the surface V in $\mathbb{P}^3$ is that a line $\ell=L_f$ cannot have any imaginable position with respect to V. For instance, if the line ℓ cuts V in two double points P and Q one cannot have $P\in W$ and $Q\in V\setminus W$ (compare to cases (d) and (d')!).

The computations of the dimensions for the α-orbits are standard using (4.17) and we omit them.

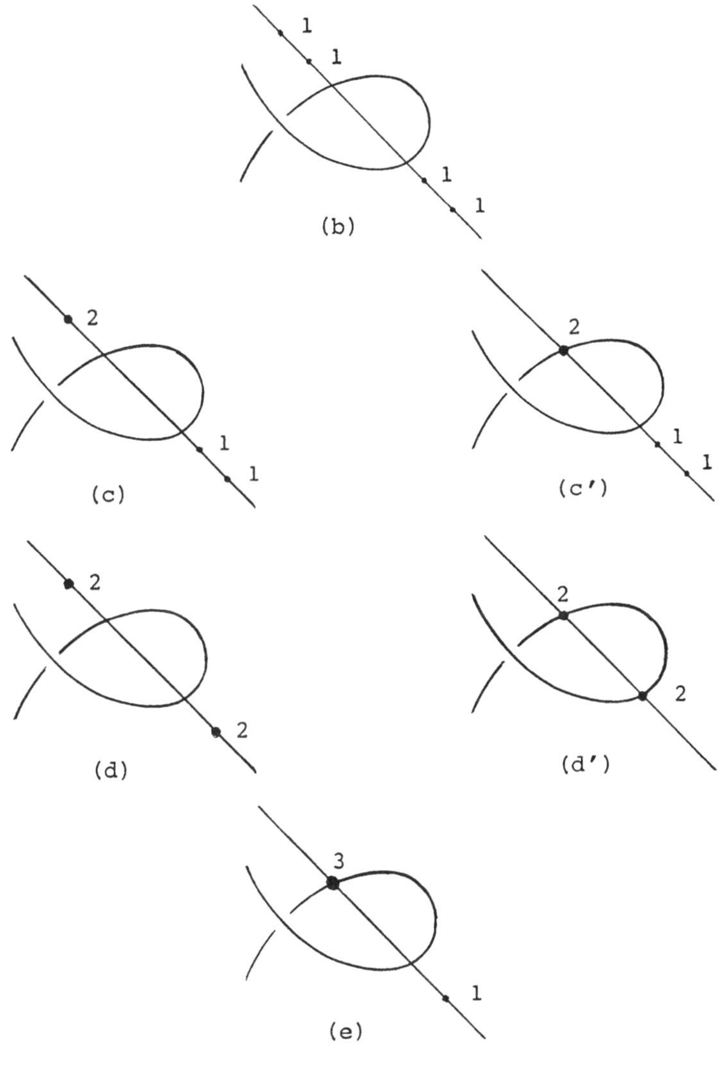

FIGURE (5.24)

(5.25) EXERCISE

Consider the normal form f_t in Table (5.22). What is the connection between the invariant $j(t)$ and the cross-ratio of the four intersection points $L_{f_t} \cap V$ (regarded as points on $L_{f_t} \simeq \mathbb{P}^1$)?

CHAPTER 6

FINITE DETERMINACY OF MAP GERMS

§1. $\mathcal{R}$-FINITE DETERMINACY

We have introduced until now two types of equivalence relations:

(a) Equivalence relations on the map germ spaces $E^o_{n,p}$ associated to some infinite dimensional actions (3.4), (3.5), (3.6) and (3.8).

(b) Equivalence relations on the jet spaces $J^k(n,p)$ associated to some algebraic actions (3.20)-(3.23). Moreover, the examples treated in Chapter 5 suggest that the listing of orbits in these jet spaces might be possible in some simple cases.

The concept of finite determinacy relates the two types of equivalence relations (a) and (b). More precisely, it implies that (for *most* map germs f) in order to determine the orbit of f in $E^o_{n,p}$ it is enough to determine the orbit of j^kf in $J^k(n,p)$ for a suitable large positive integer k.

For instance, let $f:(\mathbb{R}^n,0) \to (\mathbb{R},0)$ be a smooth function germ having a nondegenerate singularity at the origin. Since $j^1f=0$, the 2-jet j^2f is (essentially) a nondegenerate quadratic form in $H^2(n,1;\mathbb{R}) \subset J^2(n,1)$ and a normal form for its ρ-orbit (equal to its r^2-orbit) is given by the quadratic form $f_{i,r}$ from (5.5) with n=r.

Then Morse Lemma (1.3) says exactly that $f_{i,n}$ is also a normal form for the $\mathcal{R}$-orbit of f in E^o_n.

We give now the basic definitions in the context of $\mathcal{R}$-equivalence. The reader will obtain the similar notions in the other cases just by replacing $\mathcal{R}$ with $\mathcal{L}$, $\mathcal{A}$ and $\mathcal{K}$ respectively.

(6.1) DEFINITIONS

The k-th jet $j^k f$ of a map germ $f \in E^o_{n,p}$ is called *R-sufficient* if any germ $g \in E^o_{n,p}$ with $j^k g = j^k f$ is R-equivalent to the germ f. In this case we also say that the germ f is *k-R-determined*. A map germ f is called *finitely R-determined* if there is a positive integer k such that f is k-R-determined. In this case the number

$$o_R(f) = \min\{k \in \mathbb{N};\ f \text{ is k-R-determined}\}$$

is called the *R-determinacy order* of f.

(6.2) EXERCISE

Show that a jet $j^k f$ is R-sufficient if and only if any germ $g \in E^o_{n,p}$ with $j^k g \overset{R}{\sim} j^k f$ is R-equivalent to f. Similar statement for L, A, K.

In this section we discuss the simplest and historically first considered case: the R-equivalence of functions (p=1). In the next section we shall treat the K-equivalence of map germs in $E^o_{n,p}$, restricting most of the time to the complex case.

The reader which is interested in the essentially more difficult cases of L and A equivalences is suggested to consult the excellent survey [W2].

Hence from now on in this section equivalence, finitely determined and so on will mean R-equivalence, finitely R-determined and so on. Moreover, to simplify the notations, we omit the indices p=1 and n=fixed, e.g. we write m, R^k, D for m_n, R^k_n, D_n.

The results, if not stated otherwise, are valid in both real and complex cases.

(6.3) LEMMA

If the function germ $f \in m$ is k-determined, then

$TRf \supset m^{k+1}$

PROOF

Let $q=j^{k+1,k}:J^{k+1}(n,1) \to J^k(n,1)$ be the natural projection. Since f is k-determined, we have

$$q^{-1}(j^k f) \subset R^{k+1}\cdot j^{k+1}f$$

Taking the tangent spaces in both sides at the point $j^{k+1}f$ we obtain by (4.23)

$$\frac{m^{k+1}}{m^{k+2}} \subset \frac{mJ_f+m^{k+2}}{m^{k+2}}$$

Hence $m^{k+1} \subset mJ_f+m\cdot m^{k+1}$ and by Nakayama's Lemma (2.8) we get $m^{k+1} \subset m\cdot J_f = T\mathcal{R}f$.

Since the *exact converse of Lemma* (5.3) *is not true*, we have to study this situation more closely. Let $p=j^{s,k-1}:J^s(n,1) \to J^{k-1}(n,1)$ be the natural projection for $s\geq k$ and let $P^s = \ker p$. For a jet $u \in J^{k-1}(n,1)$ we introduce the *linear affine subspace* $J^s(u)=p^{-1}(u)=u+P^s$ in $J^s(n,1)$. Note that here we regard u as an element in $J^s(n,1)$ given by a polynomial function of degree $\leq k-1$.

We want to investigate the traces of the R^s-orbits on this subspace $J^s(u)$, i.e. to classify up to R^s-equivalence the s-jets having a fixed (k-1)-jet u. Let $q:R^s \to R^{k-1}$ be the natural group epimorphism, R_u^{k-1} the isotropy subgroup of u in R^{k-1} and put $G(u)=q^{-1}(R_u^{k-1})$. Then the inclusion $G(u) \subset R^s$ induces an action of the group G(u) on $J^s(n,1)$ and $J^s(u)$ is an invariant subspace.

Moreover $v_1, v_2 \in J^s(u)$ are R^s-equivalent if and only if v_1 and v_2 are G(u)-equivalent. In other words, one has the obvious equality

(6.4) $\quad R^s\cdot v \cap J^s(u) = G(u)\cdot v \quad$ for any $\quad v \in J^s(u)$.

On the other hand, the Lie algebra of the isotropy group R_u^{k-1} is equal to the kernel of the differential of the map $R^{k-1} \to J^{k-1}(n,1)$, $j^{k-1}h \mapsto j^{k-1}(u\circ h^{-1})$ at the identity element $e \in R^{k-1}$. Hence

(6.5) $$T_e \cdot R_u^{k-1} = \{ j^{k-1}X \in J^{k-1}(n,n) ; j^{k-1}(\sum_{i=1,n} X_i \frac{\partial f}{\partial x_i}) = 0 \}$$

where we assume that $u = j^{k-1}f$ for some germ $f \in m$.

This implies the following equalities

(6.6) (i) $$T_e G(u) = \{ j^s X \in J^s(n,n) ; j^{k-1}(\sum_{i=1,n} X_i \frac{\partial f}{\partial x_i}) = 0 \}$$

(ii) $$T_v(G(u) \cdot v) = T_v(R^s \cdot v) \cap P^s$$

for any jet $v \in J^s(u)$.

Note by the way that the equality (6.6. ii) would be a direct consequence of (6.4) if the manifolds $R^s v$ and $J^s(u)$ were transverse in $J^s(n,1)$. But this is not true in general!

We can now state and prove a more precise version of (6.3).

(6.7) PROPOSITION

For a function germ $f \in m$, *the next two statements are equivalent.*

(i) $TRf \supset m^k$

(ii) *The set* $U_k(f) = R^k \cdot j^k f \cap J^k(j^{k-1}f)$ *is Zariski open in* $J^k(j^{k-1}f)$.

PROOF

Since $U_k(f)$ is exactly the $G(u)$ orbit of $v = j^k f$ (where $u = j^{k-1}f$, $s = k$) by (6.4) it follows that $U_k(f)$ is Zariski open if and only if $\dim U_k(f) = \dim J^k(u)$, or $T_v U_k(f) = T_v J^k(u)$. But the last equality by (6.6.ii) is just

$$\frac{TRf \cap m^k + m^{k+1}}{m^{k+1}} = \frac{m^k}{m^{k+1}}$$

And Nakayama's Lemma shows that this is equivalent to $TRf \cap m^k \supset m^k$, hence to statement (i).

In view of this result, we introduce the next notion.

(6.8) DEFINITION

A function germ $f \in m$ is called *strongly* k-*determined* if there is a Zariski neighbourhood $U(f)$ of j^kf in $J^k(j^{k-1}f)$ such that any germ $g \in m$ with $j^kg \in U(f)$ is equivalent to f.

Some remarks are clearly in order now. First note that

(6.9) (k-1)-determined$\Rightarrow$strongly k-determined$\Rightarrow$k-determined

and, in general, no one of these implications can be reversed.

Next, in the complex case, any Zariski open set in $J^k(j^{k-1}f)$ which is not empty is dense. Hence there is at most one $G(u)$-orbit of jets j^kg with g strongly k-determined. More precisely, this unique open orbit U (if it exists) can be characterized as follows

$$U=\{j^kg \in J^k(u);\ T\mathcal{R}g \supset m^k\},\ \text{with } u=j^{k-1}f.$$

In the real case, each connected component of U might give rise to a different orbit (equivalence class). The simplest example to be kept in mind is k=2, $u=j^1f=0$ i.e. the classification of real quadratic forms (5.5). Then U is the open set of nondegenerate forms and the connected components of U correspond precisely to the various possibilities for the index $i=0,\dots,n$ of these forms (5.6). This is a *basic reason why the classification* (of jets and germs) *in the real case is more complicated, involving usually more classes than in the complex case* (most often distinguished by some $\pm$ signs).

A final remark concerns the order of determinacy. For a function germ $f \in m$ let

$$k=\min\ \{p;\ T\mathcal{R}f \supset m^p\} \leq \infty$$

and, for a finite k, we define $U_k(f)$ as in (6.7.ii).

(6.10) EXERCISE

Show that $O_{\mathcal{R}}(f)=k$ if $U_k(f) \neq J^k(u)$ and $O_{\mathcal{R}}(f)=k-1$ if $U_k(f)= =J^k(u)$, for $u=j^{k-1}f$. Note that in the complex case $U_k(f)=U$ and this gives a computable way to get $O_{\mathcal{R}}(f)$.

The next result is one of the most important ones in the study of singularities .

(6.11) THEOREM

For a function germ $f \in m$ *one has the following*
(i) $TRf \supset m^k$ *if and only if* f *is strongly* k-*determined.*
(ii) $mTRf \supset m^{k+1}$ *implies that* f *is* k-*determined.*

We give two proofs for this basic Theorem. The first proof is valid only in the complex case and is technically more sophisticated. It involves the use of Mather's Lemma (4.26) and of a simple version of Artin's Approximation Theorem (which we shall state for the reader's convenience). On the other hand, the value of this proof consists in being very clear conceptually, i.e. a step-by-step construction leading to a *formal solution* which is then approximated with an *analytic solution*. This idea is quite general and likely to be useful in many other situations as well.

The second proof is the standard one for this Theorem (see for instance [Gi], p.117). It is surprisingly simple and hence a bit misterious!

FIRST PROOF (COMPLEX CASE)

(i) First note that f strongly k-determined implies $mJ_f \supset m^k$ exactly as in the proof of (6.3). Conversely, assume now $mJ_f \supset m^k$ and take the neighbourhood U(f) from (6.8) to be $U_k(f)$. Then $U_k(f)$ is an open orbit by (6.7) and for any germ $g \in m$ with $j^k g \in U_k(f)$ there is a diffeomorphism germ $h \in D$ such that $j^k(g \circ h) = j^k f$.

Now $mJ_f \supset m^k$ clearly implies $m^2 J_f \supset m^{k+1}$. In this way, all we have to prove is in fact the second statement (ii) in the Theorem.

(ii) Recall first the groups introduced in the end of Chapter 3:

$$D(k) = \{h \in D,\ j^k h = j^k 1\} \quad \text{and} \quad D^k = D/D(k) .$$

For a positive integer $p \leq k$ we consider the quotient group $D^{k,p} = D(p)/D(k)$ which is an algebraic subgroup in the group D^k of k-jets of diffeomorphisms.

Moreover, its Lie algebra is obviously

(6.12) $\quad T_eD^{k,p}=j^k(m^pE^o_{n,n})\subset j^k(E^o_{n,n})=J^k(n,n)=T_eD^k$.

The inclusion $D^{k,p}\subset D^k$ and the r^k-action on $J^k(n,1)$ induces an action of the subgroup $D^{k,p}$ on $J^k(n,1)$ such that

(6.13) $\quad T_{j^kf}(D^{k,p}\cdot j^kf)=j^k(m^{p+1}J_f)$.

(Compare with (4.23)).

Assume now that our germ f satisfies $m^2J_f\supset m^{k+1}$ and take a germ g with $j^kg=j^kf$.

First we prove that $j^{k+1}g$ and $j^{k+1}f$ belong to the same $D^{k+1,1}$-orbit. We use Mather's Lemma (4.26) with $G=D^{k+1,1}$, $M=J^{k+1}(n,1)$ and $P=J^{k+1}(j^kf)$. By (6.13) we get

$$T_{j^{k+1}g}(G\cdot j^{k+1}g)=j^{k+1}(m^2J_g)$$

We show now that the tangent spaces $T_v(G\cdot v)$ for $v\in P$ are all equal, namely that

$$m^2J_g+m^{k+2}=m^2J_f+m^{k+2}$$

Indeed $j^kg=j^kf\Rightarrow g-f\in m^{k+1}\Rightarrow \frac{\partial g}{\partial x_i}-\frac{\partial f}{\partial x_i}\in m^k\Rightarrow m^2J_g\subset$ $\subset m^2J_f+m^{k+2}$ and the converse inclusion. Since we have $m^2J_f\supset m^{k+1}$ it follows that $T_v(G\cdot v)\supset T_vP=m^{k+1}/m^{k+2}$ for any $v\in P$.

By Mather's Lemma, $P\subset G\cdot j^{k+1}f$ and hence there is an analytic isomorphism germ $h_1\in D(1)$ such that $j^{k+1}(g\circ h_1)=j^{k+1}f$.

By induction, we can find as above a sequence of analytic isomorphism germs $h_s\in D(s)$ such that

$$j^{k+s}(g\circ h_1\circ\ldots\circ h_s)=j^{k+s}f \quad \text{for any} \quad s\geq 1.$$

From the definition of the subgroups D(s) it follows that the *infinite composition* $h=h_1\circ h_2\circ\ldots$ gives a well-defined *formal isomorphism*, i.e. the components of h are formal power series. Moreover, it is clear that $j^N(g\circ h)=j^Nf$ for any $N\geq 1$ and hence $g\circ h=f$. To end the proof, we need a device to replace the formal isomorphism h with a *convergent* (i.e. *analytic*) *isomorphism* $\bar{h}$ satisfying still the equality $g\circ\bar{h}=f$.

This can be done using a simple version of a fundamental result which we recall briefly now.

Consider two sets of indeterminates

$$X=(X_1,\dots,X_n) \quad \text{and} \quad Y=(Y_1,\dots,Y_p)$$

Let $\mathbb{C}\{X\}$, $\mathbb{C}\{X,Y\}$ (resp. $\mathbb{C}[[X]]$) denote the convergent (resp. formal) power series rings in the indeterminates X, (X,Y) respectively.

For a *system of equations* $S:F(X,Y)=0$ where $F\in\mathbb{C}\{X,Y\}^q$, $q\geq 1$ an element $Y(X)\in\mathbb{C}\{X\}^p$ (resp. $Y(X)\in\mathbb{C}[[X]]^p$) is called a *convergent* (resp. *formal*) *solution* if $Y(0)=0$ and $F(X,Y(X))=0$. For such an element $Y(X)$ its k-th jet $j^kY(X)\in J^k(n,p)$ has an obvious meaning.

(6.14) ARTIN'S APPROXIMATION THEOREM

Let $Y(X)$ *be a formal solution of the system of equations* S. *Then, for any positive integer* $k\geq 1$, *there is a convergent solution* $Y^k(X)$ *of the system* S *such that* $j^kY^k(X)=j^kY(X)$.

A proof of this result can be found for instance in [Ar] or [T], p. 59. We mention for the reader willing to work more, that in fact a weaker form of (6.14) is enough for our needs here. Namely, one can use a version of the Implicit Function Theorem known also under the name of *Newton's Lemma*, see [T], p. 57.

To end our proof, we take $p=n$, $q=1$, $F(X,Y)=f(X)-g(Y)=0$, $Y(X)=h(X)$ and $k=1$. Then there is a convergent solution $\bar{h}(X)=$ $=Y^1(X)$ of this equation and $j^1\bar{h}(X)=j^1h(X)=X$. Therefore $\bar{h}$ is indeed an analytic isomorphism in D with $g\circ\bar{h}=f$.

SECOND PROOF (REAL AND COMPLEX CASES)

As in the first proof, it is enough to show (ii).

Assume that the function germ $f\in m$ satisfies $m^2J_f\supset m^{k+1}$ and let g be a germ in m with $j^kg=j^kf$.

Then the difference $u=g-f$ is in m^{k+1} and we can consider the path (*homotopy of germs*)

$$f_t=f+tu \quad \text{for} \quad t\in K$$

Since $f_o=f$, $f_1=g$, it is enough to construct a family of germs of diffeomorphisms $h_t\in D$ such that

(6.15) $\quad f_t\circ h_t=\text{constant}, \quad \text{for} \quad t\in K.$

As K is connected, it is enough to construct the family h_t locally around a fixed point $t_o\in K$. Note that such a family h_t would generate a vector field X of the form

(6.16) $\quad X=\frac{\partial}{\partial t}+X_1(t,x)\frac{\partial}{\partial x_1}+\ldots+X_n(t,x)\frac{\partial}{\partial x_n}$

defined around $(t_o,0)$ in $K\times K^n$ with $X_k(t,0)=0$ for any $k=1,\ldots,n$ and any t. Indeed, one can take

$$X(t,x)=\frac{d}{dt}(t,h_t(x))$$

and use the equality $h_t(0)=0$.

The equality (6.15) is clearly equivalent to

(6.17) $\quad X\cdot f_t=u+\sum_{i=1,n} X_i\frac{\partial f_t}{\partial x_i}=0$

Conversely, if we have a vector field X with the properties (6.16), (6.17), one can get the family of diffeomorphisms h_t by the (local) integration of the system of ordinary differential equations associated to X. Therefore it is enough to construct such a vector field X and we do this in the following way.

Let $\bar{E}={}_{(t_o,0)}E_{n+1}$ be the K-algebra of function germs at $(t_o,0)\in K\times K^n$ and $\bar{m}$ be its maximal ideal. We note as usual in this section $E=E_n$, $m=m_n$.

The equality (6.17) is equivalent to the fact that the element u belongs to the ideal $m\cdot I$ in $\bar{E}$, where $I=(\frac{\partial f_t}{\partial x_1},\ldots,\frac{\partial f_t}{\partial x_n})$ is the ideal generated by these partial derivatives in $\bar{E}$. And $\bar{E}$ is regarded as an E-module, via the obvious inclusion $E\subset\bar{E}$ (a function of x is regarded as a function of t and x!).

Hence it is enough to prove the inclusion

$$m^{k+1}\cdot\bar{E}\subset m\cdot I$$

By Nakayama's Lemma (2.8) this is equivalent to

$$m^{k+1}\bar{E}\subset m\cdot I+\bar{m}(m^{k+1}\bar{E}) \ .$$

Let $p_1,\ldots,p_s$ be elements in m^{k+1} whose classes in m^{k+1}/m^{k+2} form a basis of this vector space. Then, by (2.9) it is enough to show that

$$p_i\in m\cdot I+\bar{m}(m^{k+1}\bar{E}) \quad \text{for} \quad i=1,\ldots,s \ .$$

By the assumption on the germ f one has

$$p_i=\sum_{j=1,n} a_{ij}\frac{\partial f}{\partial x_j} \quad \text{for} \quad i=1,\ldots,s \quad \text{and} \quad a_{ij}\in m^2 \ .$$

The elements

$$\bar{p}_i=\sum_{j=1,n} a_{ij}\frac{\partial f_t}{\partial x_j}=p_i+\sum_{j=1,n} ta_{ij}\frac{\partial u}{\partial x_j}$$

are clearly in $m\cdot I$ and $\bar{p}_i=p_i$ modulo $m^{k+2}\bar{E}\subset\bar{m}(m^{k+1}\bar{E})$.

This obviously ends the second proof of the Theorem.

It is now the time to see more concretely which function germs are finitely determined.

(6.18) EXAMPLES

(i) *Submersions*

Note that a germ $f\in m$ is a submersion germ as in (1.1) if and only if $\frac{\partial f}{\partial x_i}\notin m$ for some i. This is equivalent to $J_f=E$ and hence, in this case, one has $mJ_f=m$.

By (6.11.i) it follows that f is *strongly 1-determined.* In particular $f\sim j^1f$ which is essentially the Submersion Theorem (1.1) in the case p=1.

(ii) *Morse functions (nondegenerate singularities)*

A germ $f\in m$ has a nondegenerate singularity at the origin $0\in K^n$ as defined in (1.2) if and only if $f\in m^2$ and the classes of the partial derivatives $\frac{\partial f}{\partial x_1},\ldots,\frac{\partial f}{\partial x_n}$ in $\frac{m}{m^2}$ form a basis.

In turn this is equivalent by (2.9) to the fact that these partial derivatives generate the ideal m. In other words $J_f=m$ and hence $mJ_f=m^2$. Hence (6.11.i) implies that in this case f is *strongly* 2-*determined*. In particular $f\sim j^2f$, which is essentially Morse Lemma (1.3) as was explained at the beginning of this Chapter.

(iii) *Complex cubic forms* (n=2 *and* n=3)

Let first n=2 and f be a germ in m^3 whose 3-jet j^3f is a nondegenerate binary cubic form, i.e. up to a linear coordinate change $j^3f=x^2y+y^3$ as in Table (5.8). Then using (2.9) it is easy to see that $mJ_f=m^3$ and hence by (6.11.i) f is *strongly* 3-*determined*. In particular $f\sim x^2y+y^3$.

The next case n=3 is more interesting. Recall the normal form $f_t=x^3+y^3+z^3+3txyz$ from (5.14). Then a simple computation shows that

(a) $mJ_{f_t}\not\supset m^3$ for any $t\in\mathbb{C}$, hence f_t is not strongly 3-determined. This is clearly related to the fact that the orbits corresponding to the normal forms f_t are not open.

(b) $m^2J_{f_t}\supset m^4$ for $t\in\mathbb{C}$, $t^3\neq-1$. Hence by (6.11.ii) we get that the germ f_t is 3-*determined* if and only if the cubic curve $V(f_t)$ in $\mathbb{P}^2$ is *smooth*.

This last remark is part of a general result, which will be a corollary of the following.

(6.19) PROPOSITION

For a function germ $f\in m$, *the next statements are equivalent.*

(i) f *is finitely determined.*

(ii) $TRf\supset m^k$, *for some positive integer* k.

(iii) f *has finite codimension*, i.e.

$$\lim_{s\to\infty}\operatorname{codim}(R^s\cdot j^sf)=\dim(m/TRf)<\infty$$

In the complex case, one has in addition (iv) 0 *is at most an isolated singularity for* f, i.e.

$$V(J_f)\subset\{0\}.$$

PROOF

(i) $\Longleftrightarrow$ (ii) follows from (6.3) and (6.11).

To prove (ii) $\Longleftrightarrow$ (iii), we first note that by codim $(R^s.j^sf)$ we mean its codimension in $J^s(n,1)$. The sequence codim$(R^s.j^sf)$ is increasing and becomes stationary at the k-th term if and only if $T\mathcal{R}f \supset m^k$ (recall if necessary the proof of (2.6.iii)).

In the complex case, $V(J_f)=\emptyset$ as analytic set germs at the origin if and only if f is a submersion germ as in (6.18.i).

In general, one has $m.J_f \subset J_f$ and hence $V(mJ_f) \supset V(J_f)$. But in fact it is easy to show that for any *proper* ideal $I \subset E$ one has $V(mI)=V(I)$. Then condition (iii) is equivalent to $\dim(E/mJ_f)<\infty$ which by (2.15.ii) is equivalent to $V(mJ_f) \subset \{0\}$.

(6.20) COROLLARY

Let $f \in H^d(n,1;\mathbb{C})$ *be a nonzero homogeneous polynomial. Then* f *regarded as a function germ in* m_n *is finitely determined if and only if the projective hypersurface* $V(f) \subset \mathbb{P}^{n-1}$ *defined by* f *is smooth.*

PROOF

The case d=deg(f)=1 is trivial. Assuming $d \geq 2$, both conditions above are equivalent to the next equality of algebraic sets

$$\{x \in \mathbb{C}^n;\ \frac{\partial f}{\partial x_1}(x)=\ldots=\frac{\partial f}{\partial x_n}(x)=0\}=\{0\}.$$

One must note that the order of determinacy $O_{\mathcal{R}}(f)$ is in general strictly greater than d=deg(f) in spite of Example (6.18.iii). A precise formula for $O_{\mathcal{R}}(f)$ will be derived in the next Chapter (see (7.31)).

We end the discussion of $\mathcal{R}$-equivalence with a remark on *real analytic functions*. Let A denote the $\mathbb{R}$-algebra of germs of analytic functions $(\mathbb{R}^n,0) \to \mathbb{R}$, defined as in (2.4) with "smooth" replaced by "real analytic".

Let $E^{\mathbb{R}}$ be the $\mathbb{R}$-algebra E_n of smooth function germs $(\mathbb{R}^n,0) \to \mathbb{R}$ and $E^{\mathbb{C}}$ be the $\mathbb{C}$-algebra E_n of complex analytic function germs $(\mathbb{C}^n,0) \to \mathbb{C}$. Then one has the following inclusions and equalities

(i) $\mathbb{R}[x_1,\ldots,x_n] \subset A = \mathbb{R}\{x_1,\ldots,x_n\} \subset E^{\mathbb{R}}$

(ii) $A \subset E^{\mathbb{C}} = \mathbb{C}\{x_1,\ldots,x_n\}$

Hence a germ $f \in A$ can be regarded as a smooth germ in $E^{\mathbb{R}}$ or as a complex analytic germ in $E^{\mathbb{C}}$ called the *complexified germ associated to* f and denoted by $f^{\mathbb{C}}$.

(6.21) EXERCISE

The real analytic function germ f with $f(0)=0$ is finitely determined (as a germ in $E^{\mathbb{R}}$) if and only if the complexified germ $f^{\mathbb{C}}$ is finitely determined (as a germ in $E^{\mathbb{C}}$); and they have the same determinacy orders. Hint: use (6.19).

§2. K-FINITE DETERMINACY AND ISOLATED SINGULARITIES OF COMPLETE INTERSECTIONS

In this section we study the finite K-determinacy of map germs in $E^0_{n,p}$ and relate it (in the complex case) to an important class of germs of complex analytic spaces.

To simplify the notation, we let m denote the maximal ideal $m_n \subset E_n$ and we omit the letter K whenever this does not lead to confusion.

(6.22) LEMMA

If the map germ $f \in E^0_{n,p}$ *is* s-*determined, then*

$$TKf \supset m^{s+1} E_{n,p} .$$

PROOF

Completely similar to the proof for (6.3).

We proceed now as in the previous section.

Let $q = j^{t,s-1} : J^t(n,p) \to J^{s-1}(n,p)$ be the natural projection for $t \geq s$ and let P^t denote its kernel. For a jet $u \in J^{s-1}(n,p)$

we consider the *linear affine subspace*

$J^t(u)=q^{-1}(u)=u+P^t$ in $J^t(n,p)$.

We let q also denote the natural projection $K^t_{n,p} \to K^{s-1}_{n,p}$. Let $S(u) \subset K^{s-1}_{n,p}$ be the isotropy subgroup of u and define $G(u)=q^{-1}(S(u))$.

Then the inclusion $G(u) \subset K^t_{n,p}$ induces an action of the group G(u) on the jet space $J^t(n,p)$ and $J^t(u)$ is an invariant subspace. One has

(6.23) $\quad K^t_{n,p} \cdot v \cap J^t(u) = G(u) \cdot v$, for any $v \in J^t(u)$.

And a similar computation to the proof of (6.6) gives

(6.24) $\quad T_v(G(u) \cdot v) = T_v(K^t_{n,p} \cdot v) \cap P^t$, for any $v \in J^t(u)$.

A more precise version of Lemma (6.22) is the following

(6.25) PROPOSITION

For a map germ $f \in E^0_{n,p}$ *the next two statements are equivalent.*

(i) $TKf \supset m^s E_{n,p}$

(ii) *The set* $U_s(f) = K^s_{n,p} \cdot j^s f \cap J^s(j^{s-1}f)$ *is Zariski open in* $J^s(j^{s-1}f)$.

PROOF

Completely similar to the proof for (6.7).

In view of this result, it is natural to introduce the following notion.

(6.26) DEFINITION

A map germ $f \in E^0_{n,p}$ is called *strongly* s-*K-determined* if there is a Zariski neighbourhood U(f) of the jet $j^s f$ in $J^s(j^{s-1}f)$ such that any germ $g \in E^0_{n,p}$ with $j^s g \in U(f)$ is K-equivalent to f.

All the remarks after Definition (6.8) above apply equally well to the present case, with obvious changes e.g. $U=\{j^s g \in J^s(j^{s-1}f);\ TKg \supset m^s E_{n,p}\}$.

We can state now the basic result on finite K-determinacy, the analogue of Theorem (6.11).

(6.27) THEOREM

For a map germ $f \in E^o_{n,p}$ *one has the following.*

(i) $TKf \supset m^s E_{n,p}$ *if and only if* f *is strongly* s-K-*determined.*

(ii) $mTKf \supset m^{s+1} E_{n,p}$ *implies* f *is* s-K-*determined.*

PROOF

(i) If the map germ f is strongly s-determined, then one has $TKf \supset m^s E_{n,p}$ as in (6.22). Assume now $TKf \supset m^s E_{n,p}$ and take the neighbourhood U(f) from (6.26) to be the open G(u)-orbit $U_s(f)$.

Then for a map germ $g \in E^o_{n,p}$ with $j^s g \in U_s(f)$ one has $j^s g \sim j^s f$ via the G(u)-action, where $u = j^{s-1} f$. Since $TKf \supset m^s E_{n,p}$ obviously implies $mTKf \supset m^{s+1} E_{n,p}$, all we have to prove is (ii).

(ii) We prove this part *only in the complex case* by a method similar to that used in the first proof of Theorem (6.11). A proof in the real smooth case can be found in [Mr], Chap. XV and involves a slightly different (but equivalent!) definition of the K-equivalence relation.

First recall from Chapter 3 the definitions of the groups $D_n(s)$, $M_{n,p}(s)$ and $K^s_{n,p}$ for a positive integer $s \geq 1$.

Now for a positive integer $r \geq 1$, $r \leq s$ we define an algebraic subgroup in the contact group $K^s_{n,p}$ by the formula

$$(6.28) \qquad K^{s,r}_{n,p} = (D_n(r)/D_n(s)) \rtimes (M_{n,p}(r)/M_{n,p}(s))$$

A simple computation shows that

$$(6.29) \qquad T_{j^s g}(K^{s,r}_{n,p} \cdot j^s g) = j^s(m^r \cdot TKg))$$

for any jet $j^s g$.

We show now that the assumption in (ii) implies that all the jets in $P = J^{s+1}(j^s f)$ are in a unique $K^{s+1,1}_{n,p}$-orbit. We use for this Mather's Lemma (4.26) with $G = K^{s+1,1}_{n,p}$, $M = J^{s+1}(n,p)$ and

P as above. By (6.29) we have

$$T_v(G\cdot v)=j^{s+1}(mTKg) \quad \text{for any} \quad v=j^{s+1}g\in P$$

Hence the tangent spaces $T_v(G\cdot v)$ for $v\in P$ are all equal and $T_v(G\cdot v)\supset T_vP$.

Therefore $P\subset G\cdot j^{s+1}f$ which, in other words, means: there is an element $(h_1,A_1)\in K_{n,p}$ such that $j^{s+1}((h_1,A_1)\cdot g)=j^{s+1}f$ and $h_1\in D_n(1)$, $A_1\in M_{n,p}(1)$.

By induction we find a sequence of elements $(h_t,A_t)\in K_{n,p}$ such that $h_t\in D_n(t)$, $A_t\in M_{n,p}(t)$ and $j^{s+t}(((h_t,A_t)\cdot\ldots$ $\ldots\cdot(h_1,A_1))\cdot g)=j^{s+t}f$ for all $t\geq 1$.

Note that the limit

$$(h,A)=\lim_{t\to\infty}((h_t,A_t)\cdot\ldots\cdot(h_1,A_1))$$

is well-defined, where h is a formal isomorphism $(\mathbb{C}^n,0)\to$ $\to(\mathbb{C}^n,0)$ and A is an invertible $p\times p$ matrix with entries in $\mathbb{C}[[x_1,\ldots,x_n]]$.

Moreover, one has obviously $A\cdot(g\circ h^{-1})=f$. Using now Artin's Approximation Theorem (6.14) we get some elements $\bar{h}\in D_n$, $\bar{A}\in M_{n,p}$ such that $\bar{A}\cdot(g\circ\bar{h}^{-1})=f$. (The reader is suggested to fill in all the details e.g. to write explicitely the system S which occurs in (6.14)!). Therefore f and g are K-equivalent and this ends the proof.

(6.30) EXAMPLES

(i) Consider the map germ $f:(\mathbb{C}^n,0)\to(\mathbb{C}^2,0)$ with components $f_1=x_1^2+\ldots+x_n^2$, $f_2=a_1x_1^2+\ldots+a_nx_n^2$ where $n\geq 2$, $a_i\in\mathbb{C}$ and $a_i\neq a_j$ for $i\neq j$. This germ corresponds exactly to the *generic pencil of quadrics* from (5.20). A direct and simple computation of the tangent space TKf proves the following.

(a) $TKf\supset m^2E_{n,2}$ if and only if n=2 or n=3. In these cases the germ f is *strongly* 2-K-*determined* by (6.27.i) and, in particular, the class of f does not depend on the choice of the parameters a_i. In other words: there are no moduli in these cases.

(b) $mTKf\supset m^3E_{n,2}$ for any n and hence, by (6.27.ii) it follows that f is 2-*determined.*

(ii) Consider now the map germ $f_t:(\mathbb{C}^2,0) \to (\mathbb{C}^2,0)$ associated to the *generic pencil of binary cubic forms* $f_t(x,y)=(x^3+3x^2y,\ 3xy^2+ty^3)$, $t\in\mathbb{C}\setminus\{0,1,9\}$ from Table (5.22). Again by direct computation one shows that $mTKf_t \supset m^4\cdot E_{2,2}$ and hence, by (6.27.ii), the germ f_t is 3-determined.

We want now to investigate which map germs are finitely K-determined. Before stating the result, we have to introduce some basic notions.

Let $(X,\mathcal{O}_X)$ be an analytic space germ at the origin of $\mathbb{C}^n$. Let $f_1,\dots,f_p$ be a *minimal set of generators* for the corresponding defining ideal $I_X \subset E_n$. Then one has the following

(6.31) EXERCISE

Show that $p=\dim(I_X/mI_X)$ using (2.9).

(6.32) DEFINITION

The analytic space germ $(X,\mathcal{O}_X)$ is called a *complete intersection* if dim $X=n-p$. We also say in this situation that $(X,0)$ is a *complete intersection singularity* or a *singularity of complete intersection.*

Note that in general one has dim $X\geq n-p$ and the equality means that the function germs $f_1,\dots,f_p$ above are in some sense independent. To formalize this independency, we need the next

(6.33) DEFINITION

Let E be a ring and $a_1,\dots,a_p$ some noninvertible elements in E. The sequence $a_1,\dots,a_p$ is called a *regular sequence* in E if the class of a_j is not a zero divisor in the quotient ring $E/(a_1,\dots,a_{j-1})$ for $j=1,\dots,p$.

Now we can state the following characterization of complete intersections (see for proof [Lm], p.114):

(6.34) PROPOSITION

With the notations above, the following statements are

equivalent.

(i) (X,0) *is a complete intersection singularity.*

(ii) *The members of any minimal set of generators for the ideal* I_X *form a regular sequence in* E_n.

(iii) *The ideal* I_X *is generated by the members of a regular sequence in* E_n.

(6.35) EXAMPLES

(i) *Hypersurface singularities.* With the above notations, if the ideal I_X is principal i.e. $p=1$ then one has $\dim X=n-1$ (see for instance [Lm], p. 104). Moreover, conversely, any pure dimensional analytic space germ whose embedding codimension equals 1 is of this type [Lm], p. 105. This special class of complete intersection singularities are called *hypersurface singularities.*

(ii) *The cone over the twisted cubic* (a counterexample!).

Consider at the origin of $\mathbb{C}^4$ (coordinates a,b,c,d) the next analytic space germ

$$X: ac-b^2=ad-bc=bd-c^2=0$$

This is precisely the cone over the twisted cubic curve $W\subset\mathbb{P}^3$ considered in Chap. 5, §4.

One has obviously $\dim X=2$, but the ideal I_X cannot be generated by 2 elements (just note that the 3 defining equations of X are linearly independent!).

For any map germ $f:(K^n,0)\to(K^p,0)$ we define $I_p(f)$ to be the ideal in E_n generated by all $p\times p$-minors in the Jacobian matrix of f

$$\left(\frac{\partial f_i}{\partial x_j}\right)_{\substack{i=1,\dots,p.\\ j=1,\dots,n.}}$$

In particular, $I_p(f)=0$ for $p>n$.

Recall from Chapter 3 the notation I_f for the ideal in E_n generated by the components $f_1,\dots,f_p$ of f. For the next definition, assume that we are *in the complex case* and that the *fiber* $X=f^{-1}(0)$ *is a complete intersection of dimension* n-p. Since the analytic space germ X corresponds to the ideal I_f, this is clearly equivalent by (6.34) to the fact that $f_1,\ldots,f_p$ is a regular sequence in E_n.

(6.36) DEFINITION

With the above notations, the *singular subspace* Sing X of X is the analytic space germ at the origin of $\mathbb{C}^n$ defined by the ideal $I_f+I_p(f)$.

(6.37) EXERCISE

(i) Show that the ideal $I_f+I_p(f)$ does not depend on the choice of a minimal set of generators $f_1,\ldots,f_p$ for the ideal $I_f=I_X$ of the complete intersection X.

(ii) Show that the analytic set germ corresponding to Sing X is precisely the analytic set germ (at the origin of $\mathbb{C}^n$) of the set $S(\tilde{X})$ of singular points of a representative $\tilde{X}$ for X. Hint: use (4.4) and a fundamental algebraic result, namely the jacobian criterion for smoothness (see for instance [T], p. 33). Alternatively, use (3.15) and the remark which follows it.

(6.38) DEFINITION

Let (X,0) be a complete intersection (resp. hypersurface) singularity. If the inclusion of germs of analytic sets Sing $X \subset \{0\}$ holds, then we say that (X,0) is an *isolated complete intersection* (resp. *hypersurface*) *singularity*.

This is justified by the following remark: by (6.37.ii) if Sing $X=\{0\}$, then X admits a representative $\tilde{X}$ (let us say in a small open ball around the origin) such that $\tilde{X}\setminus\{0\}$ is smooth. Note that by definition, the class of smooth germs is contained in the class of isolated complete intersection singularities. This is convenient sometimes for unifying the statements (see (6.39) below for an example of this situation).

Our interest in the class of complete intersection isolated singularities comes from the next basic result .

(6.39) PROPOSITION

For a map germ $f\in E^o_{n,p}$ *the next statements are equivalent.*

(i) f *is finitely K-determined.*

(ii) $TKf\supset m^sE_{n,p}$ *for some positive integer* s.

(iii) f *has finite K-codimension,* i.e.

$$\lim_{s\to\infty} \operatorname{codim}(K^s_{n,p}\cdot j^sf)=\dim(E^o_{n,p}/TKf)<\infty$$

(iv) $I_f+I_p(f)\supset m^s$ *for some positive integer* s.

Moreover, in the complex case, one has in addition

(v) a. *For* $n\geq p$, *the fiber* $f^{-1}(0)$ *of the map germ* f *is an isolated complete intersection singularity of dimension* n-p.

b. *For* $n\leq p$, *one has the equality* $f^{-1}(0)=\{0\}$ *of germs of analytic sets.*

PROOF

The equivalence (i)$\Longleftrightarrow$(ii) follows from (6.22) and (6.27). Next (ii)$\Longleftrightarrow$ (iii) can be proved exactly as (2.6.iii). Namely, consider the sequence of inclusions

$$E^o_{n,p}\supset TKf+mE^o_{n,p}\supset TKf+m^2E^o_{n,p}\supset\ldots$$

By Nakayama's Lemma (2.8), this sequence becomes stationary at the s-th step if and only if

$$TKf\supset m^{s-1}E^o_{n,p}=m^sE_{n,p}\quad .$$

We mention that the sequence $\operatorname{codim}(K^s_{n,p}\cdot j^sf)$ is increasing, where codimension is taken with respect to the ambient jet space $J^s(n,p)$.

We prove now the implication (ii) $\Longrightarrow$ (iv).

Let $u=u_1\cdot\ldots\cdot u_p$ with $u_i\in m^s$ for i=1,...,p. Then there are elements $a_{ik}\in m$ such that

(6.40) $$\sum_{i=1,n} a_{ik}\frac{\partial f}{\partial x_i}=u_k\cdot e_k \quad \text{modulo}\quad I_f\cdot E_{n,p}$$

where the (column!) vectors e_k are the canonical basis of the E_n-free module $E_{n,p}$.

The equalities (6.40) for $k=1,\ldots,p$ can be written all together in matriceal form as follows

$$\left(\frac{\partial f}{\partial x_1},\ldots,\frac{\partial f}{\partial x_n}\right)\cdot(a_{ik})_{\substack{i=1,n\\k=1,p}}=(u_1e_1,\ldots,u_pe_p)$$

modulo $I_f\cdot E_{n,p}$.

It follows that u, which is the determinant of the p×p matrix in the right hand side, is a linear combination with coefficients in E_n of the p-minors in the Jacobian matrix of f. Consequently $I_p(f)+I_f\supset m^{ps}$, which is exactly (iv).

For the implication (iv)$\Longrightarrow$(ii), it is enough to prove the inclusion $J_f\supset I_p(f)\cdot E_{n,p}$. To do this, we show that for any p×p-submatrix M in the Jacobian matrix of the map germ f one has

$$(\det M)\cdot E_{n,p}\subset \langle M\rangle$$

Here (det M) denotes the principal ideal generated by det M in E_n and $\langle M\rangle$ denotes the E_n-submodule in $E_{n,p}$ generated by the columns in the matrix M.

This inclusion in turn follows from the well-known relation from Linear Algebra

$$\det M\cdot 1_p=M\cdot M^*$$

where 1_p denotes the unit p×p matrix and M^* is the matrix obtained from the matrix M by replacing each entry by its algebraic complement.

Finally we prove (iv)$\Longleftrightarrow$(v) in the complex case. Assume that (iv) holds, $n\geq p$ and let $\tilde{X}$ be a small enough representative for the germ $X=f^{-1}(0)$. Then at a point $x\in\tilde{X}\setminus\{0\}$ the Jacobian matrix of f has rank p and hence $\tilde{X}$ is (in a neighbourhood of x) a smooth submanifold of dimension n-p. This shows that dim X=n-p (i.e. X is a complete intersection) and that 0 is an isolated singularity (i.e. $\tilde{X}\setminus\{0\}$ is smooth), hence (v) a. When $p>n$ one has $I_p(f)=0$ and hence condition (iv) says exactly that $V(I_f)=\{0\}$ which is exactly (v) b.

The converse implication (v)$\Longrightarrow$(iv) follows directly from the definitions.

Now we discuss the analogue of (6.20) relating finite K-determinacy and projective Algebraic Geometry. Recall first that an algebraic variety $V \subset \mathbb{P}^{n-1}$ defined by an ideal $I_V \subset \mathbb{C}[x_1,\dots,x_n]$ is called a *(projective) complete intersection* if the ideal I_V has a system of generators $f_1,\dots,f_p$ where $p = \text{codim } V$ in the projective space $\mathbb{P}^{n-1}$.

The affine variety

$$CV = \{x \in \mathbb{C}^n;\ f_1(x) = \dots = f_p(x) = 0\}$$

is called the *(affine) cone over the variety* V and can be defined without the assumption of complete intersection (recall (6.35.ii) for a concrete example).

The germ $(CV,0)$ induced by this cone at its vertex 0 coincides to the fiber $f_V^{-1}(0)$ of the polynomial map germ $f_V = (f_1,\dots,f_p) \in E^0_{n,p}$.

(6.41) EXERCISE

Show that V is a projective complete intersection if and only if $(CV,0)$ is a complete intersection singularity. If this is the case, then V is smooth if and only if $(CV,0)$ is an isolated singularity.

(6.42) COROLLARY

The map germ f_V *is finitely* K*-determined if and only if* V *is a smooth complete intersection.*

In the end of this section we want to say a word about the connection between the finite K-determinacy and the problem of *lifting isomorphisms of analytic* $\mathbb{C}$*-algebras*.

Let f and g be complex germs in $E^0_{n,p}$. Then by (3.16) $f \overset{K}{\sim} g$ if and only if the analytic $\mathbb{C}$-algebras $Q(f) = E_n/I_f$ and $Q(g) = E_n/I_g$ are isomorphic.

Moreover, any isomorphism $h: Q(f) \to Q(g)$ can be lifted to an isomorphism $\bar{h}: E_n \to E_n$ making a commutative diagram

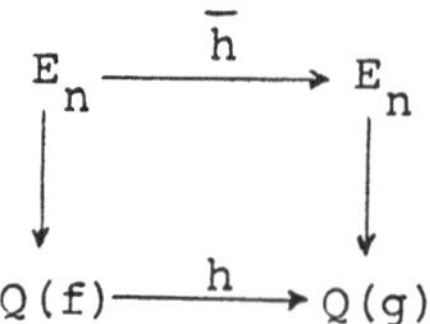

where the vertical morphisms are the canonical projections (use (i)$\Longleftrightarrow$(ii) in (3.16)).

Note that to a jet $j^k f \in J^k(n,p)$ there is associated an *artinian* $\mathbb{C}$-*algebra* $Q_k(f) = \dfrac{E_n}{I_f + m^{k+1}}$. And it is easy to show, as in the proof of (3.16), that $j^k f \overset{K}{\sim} j^k g$ if and only if the algebras $Q_k(f)$ and $Q_k(g)$ are isomorphic.

Next, the germ f is s-K-determined if and only if one has an implication:

$$Q_s(g) \simeq Q_s(f) \Longrightarrow Q(g) \simeq Q(f) \quad \text{for} \quad g \in E^o_{n,p} .$$

Hence we have again a commutative diagram (i.e. a "lifting")

$$\begin{array}{ccc} Q(f) & \xrightarrow{\;\bar{h}\;} & Q(g) \\ \downarrow & & \downarrow \\ Q_s(f) & \xrightarrow{\;h\;} & Q_s(g) \end{array}$$

but this time the vertical morphisms *can not be taken both to be the canonical projections*. More precisely, one must replace the given isomorphism h by a different one h′ which admits a true lifting as in the first diagram.

§3. GENERIC PROPERTIES OF MAP GERMS

In the beginning of this Chapter we have suggested that *most map germs are finitely determined*. In this section we are going to formalize and prove this statement, using the notion of generic properties introduced by Tougeron [T], p. 149.

First we recall the usual meaning of *a generic property P*

in Algebraic Geometry: some objects $(O_x)_{x\in X}$ are parametrized (in a natural way) by the points of an algebraic variety X and there is a Zariski open and dense subset $U\subset X$ such that $x\in U$ implies the corresponding object O_x has the property P. To have a very concrete example, let $X=H^3(3,1;\mathbb{C})\setminus\{0\}$ and for each nonzero cubic form $f\in X$, let V_f be the corresponding cubic curve in $\mathbb{P}^2$. Take U the union of all the ρ-orbits corresponding to the normal forms $x^3+y^3+z^3+3txyz$, $t\in\mathbb{C}$, $t^3\neq-1$. Then by (5.17), it follows that U is a nonempty open Zariski subset in X (and hence it is dense).

Take P the property of the cubic curve V_f to be smooth. Then by Table (5.16) it follows that V_f is smooth if and only if $f\in U$. Hence this property P is a generic property for the cubic curves (resp. for the cubic forms). One expresses this usually just by saying that a *generic cubic curve is smooth.*

The direct translation of generic properties suggested by the above considerations would be to take in the study of map germs $X=J^k(n,p)$. But since there is an open (A or K) orbit here corresponding to the nonsingular germs f with rank $df(0)=$ $=\min(n,p)$ and since any natural property P tends to be fulfilled by these germs, we would get that any such property P is generic. So, in this case, the naive point of view is not the best !

Recall that the jet spaces form an *inverse system* $(J^m(n,p), j^{m,k}$ for $m>k)$ and hence we can consider the *inverse limit*

$$J^\infty(n,p)=\varprojlim_m J^m(n,p)$$

and denote by $j^{\infty,k}:J^\infty(n,p)\to J^k(n,p)$ the canonical projections. Moreover, there is a natural map

$$T:E^o_{n,p}\to J^\infty(n,p)$$

which associates to a map germ its *Taylor series at the origin.* Note that $j^k=j^{\infty,k}\circ T$.

Assume that for each $k\geq 1$, it is given an algebraic set

$V_k \subset J^k(n,p)$ such that $j^{k+1,k}(V_{k+1}) \subset V_k$. This inclusion implies in particular that codim $V_{k+1} \geq$ codim V_k, where all the codimensions are taken with respect to the ambient jet spaces.

(6.43) DEFINITION

The inverse limit $V = \varprojlim_k V_k = \bigcap_{k\geq 1} (j^{\infty,k})^{-1}(V_k)$ is called an *algebraic provariety in* $J^\infty(n,p)$ *of codimension* equal to $\lim_k(\text{codim } V_k)$.

(6.44) LEMMA

The algebraic provariety V *above has infinite codimension if and only if for any* $u \in J^k(n,p)$ *one has* $(j^{\infty,k})^{-1}(u) \not\subset V$.

PROOF

This non inclusion is clearly a necessary condition. Conversely, let $V_k^1, \dots, V_k^s$ be the irreducible components of the algebraic set V_k. Choose $u_i \in V_k^i$ for $i=1,\dots,s$. Then the non-inclusions of $(j^{\infty,k})^{-1}(u_i)$ in V imply that there is an integer $m>k$ such that $(j^{m,k})^{-1}(u_i) \not\subset V_m$. Hence $V_m \not\supset (j^{m,k})^{-1}(V_k^i)$ for any $i=1,\dots,s$ and this obviously implies that codim $V_m >$ codim V_k.

This shows that $\lim_k (\text{codim } V_k) = \infty$, ending the proof.

(6.45) DEFINITION

A property P concerning map germs in $E^0_{n,p}$ is called *generic* if there is an algebraic provariety V in $J^\infty(n,p)$ of infinite codimension such that: for $f \in E^0_{n,p}$ the condition $T(f) \notin V$ implies that f satisfies the property P.

Note that the concept of generic in Algebraic Geometry means "*outside an algebraic variety of positive codimension*", while this concept in Singularity Theory means "*outside an algebraic provariety of infinite codimension*".

If P is a generic property, then (6.44) shows that: for any germ $f \in E^0_{n,p}$ and any positive integer $k \geq 1$, there is an approximation $g \in E^0_{n,p}$ of f of order k (in the precise sense that

$j^kg=j^kf$) such that the germ g satisfies the property P.

The main result of this section is the following

(6.46) PROPOSITION

The property of germs in $E^o_{n,p}$ *of being finitely K-determined is generic.*

PROOF

We define, for each $k\geq 1$, an algebraic subvariety V_k in $J^k(n,p)$ by the equality

$$V_k=\{j^kf\in J^k(n,p);\ \text{codim}\ (K^k_{n,p}\cdot j^kf)\geq k\}$$

The reader should check that V_k is indeed a closed algebraic variety in $J^k(n,p)$.

Now we prove that $j^kg\notin V_k$ implies that the jet j^kg is K-sufficient. For this note that

$$\text{codim}(K^k_{n,p}\cdot j^kg)=\dim \frac{E^o_{n,p}}{TKg+m^k\cdot E^o_{n,p}}$$

and consider the sequence of inclusions

$$E^o_{n,p}\supset TKg+mE^o_{n,p}\supset\ldots\supset TKg+m^k\cdot E^o_{n,p}\ .$$

If $j^kg\notin V_k$, it follows that this sequence contains at most (k-1) strict inclusions and hence

$$TKg+m^a\cdot E^o_{n,p}=TKg+m^{a+1}\cdot E^o_{n,p}$$

for some $a<k$. By Nakayama's Lemma (2.8) it follows that $TKg\supset \supset m^a\cdot E^o_{n,p}$ which shows by (6.27) that g is strongly (a+1)-K-determined.

In particular, g is k-K-determined which is what we claimed.

This shows that if we put $V=\varprojlim V_k$ then a map germ $g\in E^o_{n,p}$ with $T(g)\in J^\infty(n,p)\setminus V$ is indeed finitely K-determined.

We have still to show that codim $V=\infty$ and for this we

shall use (6.44).

Let $u \in J^k(n,p)$ and $h \in H^{k+1}(n,p;K)$ be a homogeneous polynomial map which is finitely K-determined.

The existence of such a germ h is left as an exercise for the reader. Hint: use (6.41) and the analogue of (6.21) for K-equivalence. Alternatively have a look at (7.18.iii) below.

Then the coordinate change $x_i \to tx_i$ for $t \in K\setminus\{0\}$ in D_n and multiplication by the constant scalar matrix $t^{k+1} \cdot 1_p$ in $M_{n,p}$ show that the germ g=u+h is K-equivalent to a germ $g_t(x) = u(x,t)+h(x)$ where $\lim_{t\to 0} u(x,t)=0$.

Hence the orbit $K^s_{n,p} \cdot j^s h$ is *adjacent* (or coincides) to the orbit $K^s_{n,p} \cdot j^s g$ as in (5.1). It follows that

$$m=\lim_{s\to\infty} \operatorname{codim}(K^s_{n,p} \cdot j^s g) \leq \lim_{s\to\infty} \operatorname{codim}(K^s_{n,p} \cdot j^s h) < \infty$$

and hence g is finitely determined by (6.39). Moreover, it follows that $j^r g \notin V_r$ for r=max (m+1,k+1) and hence

$$T(g) \in (j^{\infty,k})^{-1}(u)\setminus V \, .$$

We compare now the two notions of finite determinacy we have for function germs $(K^n,0) \to (K,0)$.

(6.47) PROPOSITION

A function germ $f \in m$ *is finitely R-determined if and only if* f *is finitely K-determined.*

PROOF

The inclusion of ideals $TRf \subset TKf$ shows that the implication

finite R-determinacy $\Longrightarrow$ finite K-determinacy

is trivial by (6.19) and (6.39) or just using $R \Longrightarrow K$ from (3.18).

We prove the converse implication

finite K-determinacy $\Longrightarrow$ finite R-determinacy first in the complex case, using geometric ideas. Next we derive the result in the real case from a *stronger version of the complex result*, based on a deep theorem of *Briançon-Skoda* [BS]. The idea of this argument is taken from [W2].

Assume now that f is a finitely K-determined complex analytic germ which is not finitely R-determined. By (6.19.iv) we get $\dim V(J_f)>0$ and hence there is a nonconstant curve germ $c:(\mathbb{C},0) \to (\mathbb{C}^n,0)$ such that $\operatorname{im}(c)\subset V(J_f)$. It follows that

$$\frac{\partial f}{\partial x_i}\circ c=0 \text{ for all } i \text{ and } \frac{d}{dt}(f\circ c)=\sum_{i=1,n}\left(\frac{\partial f}{\partial x_i}\circ c\right)\cdot\frac{dc_i}{dt}=0 .$$

Therefore $f\circ c=0$ and $\operatorname{im}(c)\subset V(J_f+I_f)$.

This implies that $TKf=mJ_f+I_f$ has not finite codimension in E_n, a contradiction.

We intend now to prove the next stronger result:

If $f\in m_n$ *is* s-K-*determined, then* f *is* n(s+1)-R-*determined.*

Again we start with the complex case and recall first the next

(6.48) THEOREM (BRIANÇON-SKODA)

For any function germ $f\in m_n$ *one has* $f^n\in m_n\cdot J_f$.

Now if f is s-K-determined it follows by (6.22) that

$$m^{s+1}\subset mJ_f+I_f$$

By the above theorem we get

$$m^{(s+1)n}\subset mJ_f+(f^n)\subset mJ_f$$

and hence f is n(s+1)-strongly R determined by (6.11).

We pass now to the *real smooth case*. Let $g=j^{n(s+1)}f$ and note that the complexified map germ $g^{\mathbb{C}}$ is also s-K-determined (recall (6.21)). By the stronger complex version proved above, $g^{\mathbb{C}}$ is n(s+1)-R-determined and, again by (6.21), g is n(s+1)-R-determined as a real smooth germ.

It follows that f is R-equivalent to g and hence f is

also finitely R-determined.

The above proof is a nice illustration of the basic general idea that often *the complex situation is simpler than and very helpful for understanding the real situation.*

In view of (6.47), we shall say simply about a function germ that it is finitely determined (hence omitting R or K).

§4. MILNOR NUMBER AND TJURINA NUMBER

In this section we introduce two important numerical invariants for each finitely determined function germ, in both real and complex cases. And we show that their difference represents the codimension of the leaves of a natural foliation of a K-orbit by the R-orbits contained in it, in a sufficiently large jet space.

Let $f:(K^n,0) \to (K,0)$ be a *finitely determined* function germ.

(6.49) DEFINITION

(i) The artinian K-algebra $M(f)=E_n/J_f$ is called the *Milnor algebra* of the germ f. Its dimension

$$\mu(f)=\dim_K M(f)$$

is called the *Milnor number* of the germ f.

(ii) The artinian K-algebra $T(f)=E_n/(J_f+I_f)$ is called the *Tjurina algebra* of the germ f. Its dimension

$$\tau(f)=\dim_K T(f)$$

is called the *Tjurina number* of the germ f.

(6.50) EXERCISE

Let $f,g \in m$ be finitely determined germs.

(i) If $f \overset{R}{\sim} g$, then $M(f)$ and $M(g)$ are isomorphic K-algebras. In particular $\mu(f)=\mu(g)$.

(ii) If $f \overset{K}{\sim} g$, then $T(f)$ and $T(g)$ are isomorphic K-algebras. In particular $\tau(f)=\tau(g)$.

Hint: direct computations using the definitions.

We note that the converse of (6.50.ii) is also true for complex function germs as shown by Mather-Yau [MY].

Moreover the Tjurina algebra $T(f)$ coincides to the local algebra of the analytic space germ Sing X_f as in (6.36), where X_f is the isolated hypersurface singularity defined by $f=0$. And we have conjectured and proved in some special cases in [D3] that the equivalence

$$X \sim Y \Longleftrightarrow \text{Sing } X \sim \text{Sing } Y$$

is true in the larger class of *isolated complete intersection singularities*. This problem seems to be still *open*, in spite of the substantial extensions of the Mather-Yau result due to Gaffney-Hauser [GH].

In relation with (6.50), we have also the next much more difficult result

(6.51) PROPOSITION

If $f \overset{K}{\sim} g$, *then* $\mu(f)=\mu(g)$.

PROOF

We give only some indications about this proof, since the details are too complicated for our book.

The *complex case* can be treated in two distinct ways:

(i) *topologically*, using the intepretation of $\mu(f)$ as the *degree of the gradient mapping*

$$\text{Grad}(f):(\mathbb{C}^n,0) \to (\mathbb{C}^n,0), \quad x \mapsto \left(\frac{\partial f}{\partial x_1}(x),\dots,\frac{\partial f}{\partial x_n}(x)\right)$$

see [Lm], p. 196.

Then one can use finite determinacy to reduce the problem to some jet space. Since the complex K-orbits in a jet space are connected (for details see the proof of (7.42) below) and since deg Grad(f) is a homotopy invariant, the result follows.

(ii) *algebraically*, using the interpretation of $\mu(f)$ as the *multiplicity of the ideal* J_f+I_f, see [Te], p.294.

In the *real case*, let s be the order of K-determinacy of f and g. By the proof of (6.47) we see that $f \overset{R}{\sim} \bar{f}=j^N f$, $g \overset{R}{\sim} \bar{g}=j^N g$ where $N=n(s+1)$. Next by (6.50.i), it follows that it is enough to show $\mu(\bar{f})=\mu(\bar{g})$.

Note that the complexified function germs $\bar{f}^{\mathbb{C}}$ and $\bar{g}^{\mathbb{C}}$ are K-equivalent, and hence by the complex case discussed above we have $\mu(\bar{f}^{\mathbb{C}})=\mu(\bar{g}^{\mathbb{C}})$. The result is obtained now by the simple observation that $\mu(\bar{f}^{\mathbb{C}})=\mu(\bar{f})$. See also (6.60.i) below.

Now we investigate the relation between the Milnor (resp. Tjurina) number and the R (resp. K) codimension of a function germ.

(6.52) PROPOSITION

For a finitely determined function germ $f \in m_n$ *one has the following*

(i) $\mu(f)+n=\dim(E_n/TRf)$

(ii) $\tau(f)+n=\dim(E_n/TKf)$

PROOF

(i) It is enough to show that the partial derivatives $\frac{\partial f}{\partial x_i}$ are K-linearly independent and that the sum of vector spaces

$$K\langle\frac{\partial f}{\partial x_1},\dots,\frac{\partial f}{\partial x_n}\rangle+mJ_f=J_f$$

is direct. In the contrary case, we get a relation

$$v_1\frac{\partial f}{\partial x_1}+\dots+v_n\frac{\partial f}{\partial x_n}=0 \tag{6.53}$$

with $v_i \in E$ and at least one of them not in m. Then the vector field $V=\sum_{i=1,n} v_i \frac{\partial}{\partial x_i}$ is defined on a neighbourhood of the origin and $V(0)\neq 0$.

By a standard result (see [Gi], p. 30), there is a system of coordinates $y_1,\dots,y_n$ around the origin such that $V=\frac{\partial}{\partial y_1}$.

In these coordinates, the equation (6.53) becomes $\frac{\partial f}{\partial y_1}=0$ and hence the germ f does not depend on the variable y_1. But then it is easy to see that the germ f cannot be finitely determined using (i)$\Longleftrightarrow$(ii) in (6.19).

(ii) It is enough to show that the sum

$$K<\frac{\partial f}{\partial x_1},\dots,\frac{\partial f}{\partial x_n}>+(mJ_f+I_f)=J_f+I_f$$

is direct. In the contrarycase, we get a relation

(6.54) $V(f)+a\cdot f=0$ where $a\in E$ and V is a vector field exactly as above.

Using the representation $V=\frac{\partial}{\partial y_1}$, we find an element $b\in E$ such that $V(b)=a$.

The function germ $g=\exp(b)\cdot f$ is clearly K-equivalent to f and hence it is finitely K-determined.

On the other hand one has

$$V(g)=V(\exp(b))\cdot f+\exp(b)\cdot V(f)=\exp(b)[a\cdot f+V(f)]=0$$

As above, this contradicts the finite determinacy of the germ g and ends the proof.

Let $X^s\subset J^s(n,1)$ be the $K^s_{n,1}$-orbit of a jet j^sf which is s-K-sufficient and let $g\in m$ be a function germ with $j^sg\in X^s$. Then $g\overset{K}{\sim}f$ and by (6.51) we get $\mu(g)=\mu(f)$. Since $\dim\frac{E_n}{TRg}=n+\mu(g)=$ $=n+\mu(f)$ we deduce from (2.10) and (6.11.i) that the r-jet j^rg is R-sufficient for any $r\geq\max(s,n+\mu(f))$.

If $X^r=(j^{r,s})^{-1}(X^s)=K^r_{n,1}\cdot j^rf$, it follows that all the R^r-orbits contained in X^r have the same codimension in $J^r(n,1)$ (namely $n+\mu(f)-1$) and the corresponding r-jets are all R-sufficient.

Hence these R^r-orbits give a partition of the smooth manifold X^r in *a family of smooth submanifolds of the same dimension*, i.e. *they produce a foliation on* X^r *having as leaves these submanifolds.*

Note that codim $X^r=n+\tau(f)-1$ by (6.52.ii) and all the r-jets in this $K^r_{n,1}$-orbit are K-sufficient.

We describe this situation briefly in the following statement.

(6.55) COROLLARY

The R-orbits contained in the K-orbit of a finitely determined function germ f *in a suitable large jet space* $J^r(n,1)$ *give rise to a foliation. The codimension of the leaves of this foliation is* $\mu(f)-\tau(f)$.

We end this Chapter with a more involved example of a function germ for which we compute in detail the invariants introduced so far. The reader should study it carefully, since it is a model for many other similar computations.

(6.56) EXAMPLE

Consider the family of function germs

$f_t:(K^2,0)\to(K,0)$, $f_t(x,y)=x^2y^2+t(x^5+y^5)$ for $t\in K\setminus\{0\}$.

First we prove that any germ f_t in this family is finitely determined. By (6.21) and (6.47) it is enough to treat the complex case.

Then by Proposition (6.19) it is sufficient to show that $V(J_{f_t})=\{0\}$. Indeed, the system

$$\begin{cases} f_{t,x}=\dfrac{\partial f_t}{\partial x}=2xy^2+5tx^4=0 \\ \\ f_{t,y}=\dfrac{\partial f_t}{\partial y}=2x^2y+5ty^4=0 \end{cases}$$

has only the trivial solution $(x,y)=(0,0)$ in a suitable small neighbourhood of the origin in $\mathbb{C}^2$.

Next we compute the Jacobian ideal J_{f_t}.

The most convenient way to do this in our case (and, in general, when *we are dealing with a function* f *in two variables* x,y *such that the partial derivatives* f_x,f_y *involve each at most two monomials*) is to use the following *graphical device* [AGV], p.213.

We represent the monomial x^ay^b by the integral point (a,b) in the first quadrant $\mathbb{R}^2_+$ for all $a,b\geq 0$. Then we represent each partial derivative $f_{t,x}$ and $f_{t,y}$ by the line segment joining the two points corresponding to the two monomials which enter in this partial derivative.

Any segment which is obtained from one of these two segments by a translation given by a vector $v=(v_1,v_2)$, with v_1,v_2 positive integers with $v_1+v_2>0$, is called an *admissible segment*.

The key observation is that a monomial $x^a y^b$ belongs to mJ_{f_t} if one can find a sequence of admissible segments forming a broken line, starting at the point (a,b) and ending at a point (A,B) which is as far as we want from the origin (0,0). Indeed, since f_t is finitely determined, such a point (A,B) corresponds to a monomial $x^A y^B \in mJ_{f_t}$. Then, going backwards along the broken line, we get $x^a y^b \in mJ_{f_t}$. In the next picture (6.57) we have drawn such a broken line starting from the point (6,0) corresponding to the monomial x^6 and obviously going infinitely far from the origin.

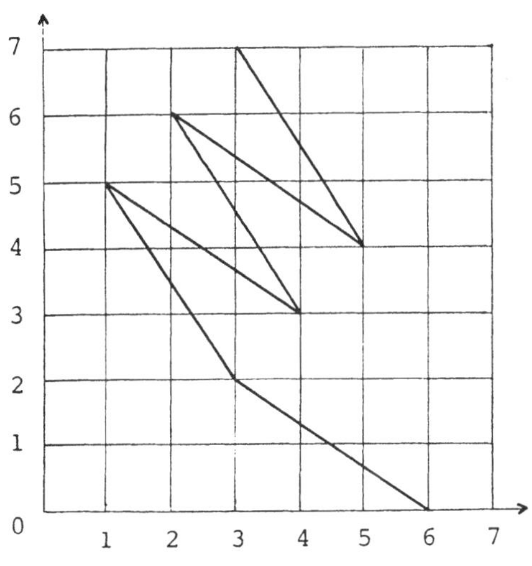

(6.57) FIGURE

The reader can easily prove, using a case-by-case analysis, that for any monomial $x^a y^b$ with a+b=6 such a broken line can be found.

Then by Nakayama's Lemma (2.8) we infer that

$$mJ_{f_t} \supset m^6$$

More precisely, since all the admissible segments used to construct these seven broken lines are obtained by translations with vectors $v=(v_1,v_2)$ with $v_1+v_2\geq 2$, we have in fact

$$m^2 J_{f_t} \supset m^6$$

By Theorem (6.11.ii), each function germ f_t is 5-$\mathcal{R}$-*determined.*

Next a simple direct computation using truncation at degree 6 shows that

$$mJ_{f_t} = K<2x^2y^2+5tx^5,\ 2x^2y^2+5ty^5,\ xy^3,\ x^3y,\ xy^4,\ x^2y^3,\ x^3y^2,\ x^4y>+m^6 .$$

Moreover, as in the proof of (6.52), one has a direct sum

$$J_{f_t} = K<f_{t,x},\ f_{t,y}>+mJ_{f_t} .$$

We can represent the situation graphically by Figure (6.58).

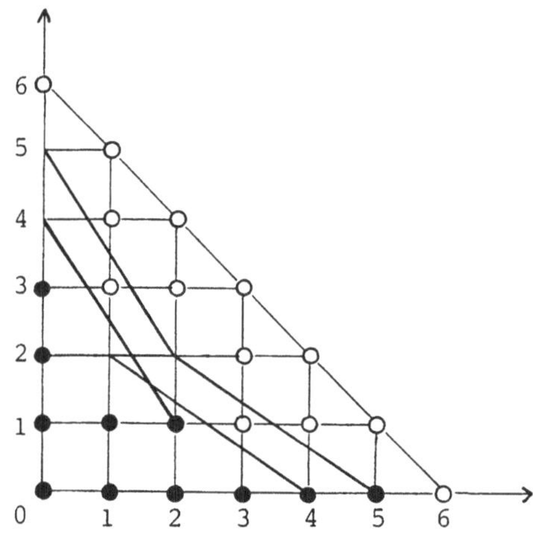

(6.58) FIGURE

Here the white circles o correspond to monomials in the jacobian ideal J_{f_t} and the four segments correspond to the binomials $f_{t,x}$, $f_{t,y}$, $xf_{t,x}$ and $yf_{t,y}$. Using this picture, it is easy to show that a K-vector space basis for the Milnor algebra $M(f_t)$ is given by the monomials corresponding to the points marked by black circles .Note only that two points joined by a segment correspond to two monomials equal (up to multiplication by constants) modulo J_{f_t}.

In particular, we get $\mu(f_t)=11$ for all $t\in K\setminus\{0\}$.

Now we pass to the invariants of the function germs f_t related to K-equivalence. The explicit description of the ideals mJ_{f_t} and J_{f_t} given above implies:

(6.59) (i) $f_t\notin J_{f_t}$ and

(ii) $TKf_t=mJ_{f_t}+K<f_t>\supset m^5$ for any $t\in K\setminus\{0\}$.

The last inclusion shows that the germ f_t is *strongly 5-K-determined*. In particular $f_t\overset{K}{\sim}f_s$ for s sufficiently close to t. In the complex case, since $\mathbb{C}\setminus\{0\}$ is connected, this implies $f_t\overset{K}{\sim}f_1$, i.e. all the function germs f_t are in a single K-orbit and a normal form for this orbit is

$$f_1=x^2y^2+x^5+y^5$$

In the real case, $\mathbb{R}\setminus\{0\}$ has two connected components and we get $f_t\overset{K}{\sim}f_1$ for $t>0$, $f_t\overset{K}{\sim}f_{-1}$ for $t<0$. But the coordinate change $x\longmapsto -x$, $y\longmapsto -y$ shows that $f_1\overset{K}{\sim}f_{-1}$, i.e. we have again a single K-orbit with normal form f_1. From (6.59) it follows that a K-vector space basis for the Tjurina algebra $T(f_1)$ is given by the basis of the Milnor algebra $M(f_1)$ described above from which we delete the monomial x^5.

Hence $\tau(f_1)=10$.

As to the orders of determinacy, we clearly get

$$O_R(f_t)=O_K(f_1)=5 \quad \text{for any} \quad t\in K\setminus\{0\} .$$

For any $s\geq 5$, the $K^s_{2,1}$-orbit of j^sf_1 has a *codimension one foliation* given by the R^s-orbits $R^s\cdot j^sf_t$, $t\in K\setminus\{0\}$. This follows from (6.55) using $\mu(f_1)-\tau(f_1)=1$.

This shows that we have a *modulus* for the family of R-orbits of the germs f_t. Now we determine a function $j:K\setminus\{0\}\to K$ such that: $f_t\overset{R}{\sim}f_s\Leftrightarrow j(t)=j(s)$. Such a function j can be called a *basic invariant for the family* f_t (compare to (5.15), (5.22) and further examples in [DG3]).

Assume that $f_s = f_t \circ h$ for some diffeomorphism $h \in D_2$. The corresponding equality at the 4-jet level and the symmetry in x,y of the situation show that h must be of the form

$$\bar{x} = h_1(x,y) = ax+p, \quad \bar{y} = h_2(x,y) = a^{-1}y+q$$

where $a \in K\setminus\{0\}$ and $p,q \in m^2$.

This gives

$$f_s(x,y) = f_t(\bar{x},\bar{y}) = x^2y^2+2ax^2yq+2a^{-1}xy^2p+t(a^5x^5+a^{-5}y^5)$$
$$\text{modulo } m^6.$$

This in turn implies $a^5 = a^{-5}$ and hence $a^5 = \pm 1$. As a result $s = \pm t$ and therefore we can take $j(t) = t^2$ as a basic invariant for the family f_t with respect to the $\mathcal{R}$-classification.

(6.60) EXERCISE

(i) Show that for a finitely determined real analytic function germ f one has:

$\mu(f) = \mu(f^{\mathbb{C}})$ and $\tau(f) = \tau(f^{\mathbb{C}})$. Compare to (6.21).

(ii) Show that a function germ f such that all its jets $j^k f$ for $k \geq k_o$ are finitely determined, is not necessarily finitely determined. Hint: take $f = (x+\frac{y}{1-y})^2$.

CHAPTER 7

WEIGHTED HOMOGENEOUS SINGULARITIES

§1. WEIGHTED HOMOGENEOUS POLYNOMIALS AND $\mathbb{C}^*$-ACTIONS

The results in the previous Chapter reduce the study of most (more precisely of finitely determined) function and map germs f to the study of the corresponding *polynomial functions and mappings* $j^k f$ given by a suitable high jet.

But even the study of a polynomial function can be very difficult, as suggested by our Example (6.56) and, in fact, by the whole Algebraic Geometry.

That is why in this Chapter we introduce a special class of polynomial mappings, whose properties are rather well understood at present and which plays an important role in the study of singularities.

Let $S=K[x_1,\dots,x_n]$ be the ring of polynomials in n indeterminates $x_1,\dots,x_n$ with coefficients in $K=\mathbb{R},\mathbb{C}$. We give each indeterminate x_i a *weight* $w_i=\mathrm{wt}(x_i)$ which is a strictly positive integer and denote by $\underline{w}=(w_1,\dots,w_n)$ the ordered set of weights.

(7.1) DEFINITION

A polynomial $f=\sum_{\underline{a}\in I} c_{\underline{a}}\cdot x^{\underline{a}}$ in S is called *weighted homogeneous* (or *quasi-homogeneous*) *of degree* d *with respect to the weights* $\underline{w}$ *if*

$$\underline{w}\cdot\underline{a}=w_1a_1+\dots+w_na_n=d$$

for any multiindex $\underline{a}\in I\subset\mathbb{N}^n$.

We use the notation $d=\deg_{\underline{w}}(f)$ and usually omit to mention the weights $\underline{w}$ when they are clear from the context. A weighted homogeneous polynomial f as above is called of *type* $(\underline{w},d)$. In this way, one gets a *grading* $S=\bigoplus_{d\geq 0} S_d$ of the polynomial ring with S_d equal to the K-vector space of weighted homogeneous polynomials in S of degree d (with respect to $\underline{w}$). Note that it is possible to have $S_d=0$ for some $d>0$.

(7.2) REMARK

Some people prefer to use positive *rational* weights $w_i'=w_i\cdot d^{-1}$ in order to insure deg(f)=1 with respect to these new weights (see for example [AGV] p.192). Sometimes it is useful to *normalize* the weights $\underline{w}$ in the following way. Let $w_o=\text{G.C.D.}\{w_1,\dots,w_n\}$, where G.C.D.=*greatest common divisor*.

Then the integers $w_i'=w_i\cdot w_o^{-1}$ are relatively prime, i.e. G.C.D. $\{w_1',\dots,w_n'\}=1$ and any weighted homogeneous polynomial f in the graded ring $S = K[x_1,\dots,x_n]$ which is of degree d with respect to $\underline{w}$ is weighted homogeneous of degree $d\cdot w_o^{-1}$ with respect to $\underline{w}'=(w_1',\dots,w_n')$. This shows that we *can always assume that the weights are relatively prime.*

(7.3) EXAMPLES

(i) The *homogeneous* polynomials in $H^d(n,1;K)$ are weighted homogeneous of degree d with respect to the weights $w_1=\dots=w_n=1$.

(ii) A polynomial of the form

$$f=x_1^{a_1}+\dots+x_n^{a_n} \text{ with } a_i\geq 2$$

is called a *Brieskorn-Pham polynomial.* If we put $d=a_1\cdot\ldots\cdot a_n$, $w_i=d\cdot a_i^{-1}$ then f is weighted homogeneous of degree d with respect to the weights $\underline{w}=(w_1,\dots,w_n)$.

(iii) Consider the polynomial $f=x^2y+y^k+z^2$, $k\geq 3$. In order to decide if this f is weighted homogeneous, one has to find the integer strictly positive solutions of the system: $d=2w_1+w_2=kw_2=2w_3$. One can take: $w_3=k$, $w_2=2$, $w_1=k-1$ and $d=2k$ and this shows that f is indeed weighted homogeneous.

Note that the number of equations involved in such a system for an arbitrary polynomial g is essentially one less than the number of monomials in g. *Hence the chances for a polynomial* g *to be weighted homogeneous are greater as the number of monomials in* g *is smaller.*

For any system of weights $\underline{w}=(w_1,\ldots,w_n)$ there is an associated *linear action of the multiplicative group* $K^*=K\setminus\{0\}$ *on* K^n given by the formula

(7.4) $\quad t\cdot x=t\cdot(x_1,\ldots,x_n)=(t^{w_1}x_1,\ldots,t^{w_n}x_n)$.

(7.5) EXERCISE

(i) Show that in the complex case this action is *effective* (i.e. $t\cdot x=x$ for all $x\in\mathbb{C}^n$ implies $t=1$) if and only if the weights $(w_1,\ldots,w_n)$ are relatively prime. Compare to (7.2).

(ii) A polynomial $f\in S$ is weighted homogeneous of degree d with respect to the weights $\underline{w}$ if and only if f satisfies the equality

$$f(t\cdot x)=t^d f(x) \quad \text{for all } x\in K^n, \quad t\in K^*.$$

If we differentiate this equality with respect to t and then put t=1 we get the famous

(7.6) EULER FORMULA

$$\sum_{i=1,n} w_i x_i \frac{\partial f}{\partial x_i}=d\cdot f$$

(7.7) EXERCISE

Let f be a weighted homogeneous polynomial and consider the following affine hypersurfaces

$$H_t=\{x\in K^n;\ f(x)=t\} \quad \text{for } t\in K^*$$

Show that H_t is either empty or a smooth hypersurface in K^n (in the complex case only the latter case is possible!).

When two such hypersurfaces H_t and H_s are isomorphic ?

In the complex case, one has as usual more results connecting Algebra with Geometry. Recall that to a complex polynomial $f\in S$ corresponds an affine algebraic set (reduced hypersurface):

$$V(f)=\{x\in\mathbb{C}^n;\ f(x)=0\}$$

(7.8) PROPOSITION

The polynomial f *is weighted homogeneous with respect to the weights* $\underline{w}$ *if and only if the algebraic set* V(f) *is invariant with respect to the* $\mathbb{C}^*$*-action* (7.4), i.e. $t\in\mathbb{C}^*$, $x\in V(f)$ *imply* $t\cdot x\in V(f)$.

PROOF

If we assume the polynomial f to be weighted homogeneous, then everything is clear by (7.5.ii).

Conversely, assume that the set V(f) is $\mathbb{C}^*$-invariant. Let $f=g_1^{a_1}\dots g_p^{a_p}$ be the decomposition of f into irreducible factors in S. Then it is clearly enough to show that each polynomial g_i is weighted homogeneous with respect to $\underline{w}$.

Note that $V_i=V(g_i)$ for $i=1,\dots,p$ are exactly the irreducible components of the hypersurface V(f). The image of the algebraic map

$$\mathbb{C}^*\times V_i \to V(f)$$

induced by the action (7.4) is irreducible and contains (and hence must coincide to) the irreducible component V_i.

This shows that V_i is a $\mathbb{C}^*$-invariant set itself, for any $i=1,\dots,p$.

In other words, we may and do assume from now on that the polynomial f is irreducible. Let $f=f_a+f_{a+1}+\dots+f_b$ be the expansion of f as a sum of weighted homogeneous components $f_k\in S_k$, with $f_a,f_b\neq 0$. For $x\in V(f)$ one has:

$$0=f(t\cdot x)=\sum_{k=a,b} t^k f_k(x) \quad \text{for all} \quad t\in\mathbb{C}^*$$

Hence $f_k\in I(V(f))$ and by *Hilbert Nullstellensatz for the polynomial ring* S (same statement as (2.14)) one gets

$$I(V(f))=\sqrt{(f)}=(f) \ .$$

The last equality follows from the fact that f is an irreducible polynomial, i.e. the principal ideal (f) in S is prime.

In particular we get $f_a=f\cdot g$ for some polynomial $g\in S$. Analyzing the weighted homogeneous components in both sides of this equality, it follows that the only possibility is $f=f_a$ an $g=1$. Therefore f is weighted homogeneous.

There is also a local analytic result similar to this global algebraic one. First we need a new definition.

(7.9) DEFINITION

The germ of an analytic set $(X,0)\subset(\mathbb{C}^n,0)$ (and the corresponding reduced analytic space germ or singularity) is called *weighted homogeneous* with respect to the weights $\underline{w}$ if the germ $(X,0)$ is $\mathbb{C}^*$-invariant with respect to the action (7.4).

The last condition means more precisely the following: there is an open ball B centered at the origin of $\mathbb{C}^n$ and a representative $\tilde{X}$ of the germ $(X,0)$ in the ball B such that $t\cdot x\in\tilde{X}$ for all $x\in\tilde{X}$ and $t\in\mathbb{C}^*$, $|t|\leq 1$.

(7.10) PROPOSITION

The analytic set germ $(X,0)\subset(\mathbb{C}^n,0)$ *is weighted homogeneous with respect to the weights* $\underline{w}$ *if and only if its ideal* $I=I(X)\subset E_n$ *is generated by weighted homogeneous polynomials with respect to the weights* $\underline{w}$.

PROOF

If the ideal I is generated by the weighted homogeneous polynomials $g_1,\dots,g_p$ then we can take as B any open ball

$$B=\{x\in\mathbb{C}^n;\ |x_1|^2+\dots+|x_n|^2<r\}$$

and as a representative

$$\tilde{X}=\{x\in B;\ g_1(x)=\dots=g_p(x)=0\}$$

Note that $|t|\leq 1$, $x\in B$ imply $t\cdot x\in B$.

Conversely, assume now that the singularity $(X,0)$ is weighted homogeneous. By Hilbert Nullstellensatz (2.14) we know that the ideal I coincides with its radical $\sqrt{I}$.

Let $f_1,\dots,f_p\in E_n$ be a system of generators for I, B an open ball centered at the origin on which all the germs f_i are defined and

$$\tilde{X}=\{x\in B;\ f_1(x)=\dots=f_p(x)=0\}$$

a representative for $(X,0)$ which is $\mathbb{C}^*$-invariant. Then for

$x \in X$ and $t \in \mathbb{C}^*$, $|t| \leq 1$ one has

(7.11) $$0=f_i(t\cdot x)=\sum_{d>0} t^d f_{i,d}(x)$$

Here $f_{i,d}$ denotes the weighted homogeneous component of f_i of degree d with respect to $\underline{w}$, i.e. one has an *infinite sum decomposition* $f_i=\sum_{d>0} f_{i,d}$. Then (7.11) implies that $f_{i,d} \in I$ for all i and d. Let J denote the ideal in E_n generated by all these weighted homogeneous polynomials $f_{i,d}$. If follows that

$$I+m_n^k=J+m_n^k \quad \text{for all} \quad k\geq 1$$

To end the proof, we need the next basic result from Commutative Algebra [AM].

(7.12) THEOREM (KRULL)

Let E *be a local noetherian ring,* m *its maximal ideal and* M *a finitely generated* E*-module. Then*

$$\bigcap_{k\geq 1} (m^k\cdot M)=0$$

In particular, for any ideal $I \subset E$ *one has*

$$\bigcap_{k\geq 1} (I+m^k)=I$$

(Just apply the Theorem to the case M=E/I.)

Hence, coming back to our proof, I=J and therefore I is generated by weighted homogeneous polynomials. Since E_n is noetherian, any set of generators for an ideal $I \subset E_n$ contains a finite subset which still generates the ideal I.

(7.13) EXERCISE

Assume in (7.10) that (X,0) is a complete intersection singularity, i.e. the ideal I is generated by p=n-dim X elements in E_n. Show that I is generated by p weighted homogeneous polynomials if we assume that (X,0) is weighted homogeneous.

(7.14) REMARK

In the real case, the example

$$V(x_1^2+\ldots+x_n^2)=\{0\}$$

shows that one cannot hope to obtain results similar to (7.8), (7.10).

§2. FINITELY DETERMINED WEIGHTED HOMOGENEOUS SINGULARITIES

We consider now the problem of the existence of finitely K-determined map germs defined by weighted homogeneous polynomials. In the case of function germs, this contains by (6.47) the problem of finite R-determinacy.

Let $H(\underline{w},\underline{d};K)$ denote the K-vector space of polynomial mappings $f:(K^n,0) \to (K^p,0)$ whose components f_i are weighted homogeneous polynomials of degree d_i with respect to the weights $\underline{w}$, for $i=1,\dots,p$. Here $\underline{d}=(d_1,\dots,d_p)$ is called the *multidegree of the map* f.

When p=1, we simply write $H(\underline{w},d_1;K)$.

(7.15) PROPOSITION

The set $A \subset H(\underline{w},\underline{d};K)$ *of polynomial mapping germs which are not finitely* K*-determined is a closed algebraic subset.*

PROOF

We treat first the complex case $K=\mathbb{C}$. In analogy to the usual projective space $\mathbb{P}^{n-1}=\mathbb{P}(\mathbb{C}^n)$, one can define the *weighted projective space* $\mathbb{P}(\underline{w})=(\mathbb{C}^n\setminus\{0\})/\mathbb{C}^*$ where the quotient is taken with respect to the $\mathbb{C}^*$-action (7.4). Equivalently, and more in the style of modern Algebraic Geometry, one can define

$$\mathbb{P}(\underline{w})=\mathrm{Proj}(\mathbb{C}[x_1,\dots,x_n])$$

where the polynomial ring is graded by the weights $w_i=\mathrm{wt}(x_i)$. It is known that $\mathbb{P}(\underline{w})$ is a *complete* algebraic variety, in general with singularities, see for instance [Do].

The set

$$B=\{(x,f)\in \mathbb{P}(\underline{w})\times H(\underline{w},\underline{d};\mathbb{C})\,;\, f(x)=0,\ \mathrm{rank}(\frac{\partial f_i}{\partial x_j}(x))<p\}$$

where $i=1,\dots,p$; $j=1,\dots,n$ is clearly a closed algebraic set.

From the completeness of the weighted projective space, it follows that $q(B)$ is a closed algebraic set in $H(\underline{w},\underline{d};\mathbb{C})$, where $q\colon \mathbb{P}(\underline{w})\times H(\underline{w},\underline{d};\mathbb{C}) \to H(\underline{w},\underline{d};\mathbb{C})$ denotes the projection (see [Mu], p.33 to get an idea of this *Elimination Theory* type result!).

Use now (6.39) (i) $\Longleftrightarrow$ (v) to show that $q(B)=A$ and hence to end the proof in this case.

Consider now the real case. To simplify the notations, we put $H^K=H(\underline{w},\underline{d};K)$ and

$$A^K=\{f\in H^K;\ f \text{ is not finitely } K\text{-determined}\} \quad \text{for } K=\mathbb{R},\ \mathbb{C}.$$

We have an *involution* $\iota\colon H^{\mathbb{C}} \to H^{\mathbb{C}}$ coming from the complex conjugation of the coefficients and it is easy to see that

(a) $H^{\mathbb{R}}=\{f\in H^{\mathbb{C}};\ \iota(f)=f\}$

(b) $f\in A^{\mathbb{C}}$ if and only if $\iota(f)\in A^{\mathbb{C}}$ (use 6.39)).

Hence the algebraic set $A^{\mathbb{C}}$ is ι-invariant and its ideal $I(A^{\mathbb{C}})$ can be generated by some polynomials $g_1,\dots,g_m$ which are ι-invariant. In other words, g_j are real polynomials and it is clear that $A^{\mathbb{R}}$ coincides to $A^{\mathbb{C}}\cap H^{\mathbb{R}}$ (compare to (6.21)) and is defined in $H^{\mathbb{R}}$ by the equations $g_1=\dots=g_m=0$.

(7.16) COROLLARY

The set $A\subset H(\underline{w},d;K)$ *of polynomial function germs which are not finitely R-determined is a closed algebraic subset.*

PROOF

Use (7.15) and (6.47).

Whether or not the set A equals the whole space $H(\underline{w},\underline{d};K)$ depends on the weights $\underline{w}$ and the multidegree $\underline{d}$.

(7.17) EXERCISE

Show that the proof of (7.15) implies that

$$A^{\mathbb{R}}=H^{\mathbb{R}} \text{ if and only if } A^{\mathbb{C}}=H^{\mathbb{C}}\ .$$

(7.18) EXAMPLES

(i) Let $p=1$, $n=2$, $\underline{w}=(2,3)$ and $d=7$. Then

$H(\underline{w},d;K)=K<x^2y>=A$

(ii) Let now consider in detail what weights $\underline{w}$ and degrees d are possible in the case p=1, n=2 such that $A\neq H(\underline{w},d;K)$ (compare to [AGV], p. 217). The basic remark is that a polynomial f belongs to A if it is divisible by x^2 or y^2. And that the weighted homogeneity type $(\underline{w},d)$ of f is determined by any two monomials with nonzero coefficients in f (a line is determined by two points!).

There are two main cases.

(A) f *contains the monomial* xy. Then either f is homogeneous of degree 2 i.e. $\underline{w}=(1,1)$, d=2 or f is of the form $axy+by^k$ with $a,b\neq 0$ and then $\underline{w}=(k-1,1)$, d=k (and a symmetric possibility obtained by $x\to y$, $y\to x$). In all these cases it is clear that $A\neq H(\underline{w},d;K)$.

(B) f *does not contain the monomial* xy. Then using the fact that x^2 and y^2 do not divide f, we get the following possibilities (modulo the symmetry in x, y).

(7.19) TABLE

Monomials in f	w_1, w_2	d	Conditions
x^a, y^b	b, a	ab	$a\geq 1$, $b\geq 1$
x^ay, y^b	b-1, a	ab	$a\geq 1$, $b\geq 2$
x^ay, xy^b	b-1, a-1	ab-1	$a\geq 2$, $b\geq 2$

Here $w_1=wt(x)$, $w_2=wt(y)$. Note that f may contain some other monomials besides those listed in the first column and that these 3 cases are not disjoint. Moreover, the possibilities from case (A) are contained in the second row of Table (7.19) for a=1.

A similar discussion in the case n=3 is possible and can be found in [AGV], p. 220.

(iii) Let $a_i\geq 2$ be positive integers for i=1,...,n, $d=a_1\ldots a_n$ and $w_i=d\cdot a_i^{-1}$. Define the weights $\underline{w}=(w_1,\ldots,w_n)$,

$\underline{d}=(d,\ldots,d)$ p-times and a mapping $f\in H(\underline{w},\underline{d};K)$ by

$$f_j=b_{j1}x_1^{a_1}+\ldots+b_{jn}x_n^{a_n} \quad \text{for} \quad j=1,\ldots,p .$$

We leave as an *exercise* for the reader to check that $f\notin A$ if the coefficients $b_{jk}\in K$ are chosen sufficiently general. (When $p\leq n$, the corresponding isolated complete intersection singularity $X=f^{-1}(0)$ is called of *Brieskorn-Pham type*.)

Our next aim is to find an *explicit formula for the Milnor number* $\mu(f)$ of a function germ $f\in H(\underline{w},d;K)\setminus A$, in terms of the weights $\underline{w}$ and the degree d (i.e. independent of the coefficients of f!).

First we recall some basic facts on *graded* K-*algebras of finite type*. For such an algebra

$$R=\bigoplus_{s\geq s_0} R_s \quad , \quad \text{with} \quad s,s_0\in\mathbb{Z}$$

the corresponding *Poincaré series* $P_R(t)$ is the Laurent formal series defined by the formula

$$(7.20) \qquad P_R(t)=\sum_{s\geq s_0}(\dim_K R_s)t^s$$

(7.21) EXAMPLE

Let R=S, the polynomial algebra $K[x_1,\ldots x_n]$ graded according to the weights $w_i=wt(x_i)$ for $i=1,\ldots,n$. Then, a simple use of the formula

$$\frac{1}{1-t^w}=1+t^w+t^{2w}+\ldots \quad \text{shows that}$$

$$P_S(t)=\frac{1}{(1-t^{w_1})\ldots(1-t^{w_n})}$$

More generally, one has the next basic result.

(7.22) PROPOSITION

Assume that $f_1,\ldots,f_p\in S$ *are weighted homogeneous polynomials of degrees* $d_1,\ldots,d_p$ *with respect to* $\underline{w}$ *and that* $f_1,\ldots,f_p$

form a regular sequence in S. *Then the quotient algebra* $R=S/(f_1,\dots,f_p)$ *is a graded* K*-algebra of finite type in a natural way and its Poincaré series is*

$$P_R(t)=\frac{(1-t^{d_1})\dots(1-t^{d_p})}{(1-t^{w_1})\dots(1-t^{w_n})}$$

PROOF

Consider the quotient algebras

$$S^o=S,\quad S^i=S/(f_1,\dots,f_i)\quad \text{for}\quad i=1,\dots,p\ .$$

Then $f_1,\dots,f_p$ is a regular sequence in S if and only if the sequences

$$0\rightarrow S^{i-1}(-d_i)\xrightarrow{f_i} S^{i-1}\longrightarrow S^i\longrightarrow 0$$

for $i=1,\dots,p$ are all exact. Here the morphism f_i is just multiplication by the polynomial f_i. And for any graded algebra T and integer $m\in\mathbb{Z}$ we define a new *shifted* graded algebra T(m) by the relation $T(m)_s=T_{m+s}$.

For the corresponding Poincaré series one clearly has

$$P_{T(m)}(t)=t^{-m}P_T(t)\ .$$

Then the exact sequences above imply

$$P_{S^{i-1}}(t)=P_{S^i}(t)+t^{d_i}\ P_{S^{i-1}}(t)$$

and this obviously ends the proof.

We give now a geometric interpretation of the fact that some weighted homogeneous polynomials $f_1,\dots,f_p$ in $\mathbb{C}[x_1,\dots,x_n]$ form a regular sequence. The reader should compare this result to (6.34).

(7.23) PROPOSITION

A sequence $f_1,\dots,f_p$ *of weighted homogeneous polynomials in* $S=\mathbb{C}[x_1,\dots,x_n]$ *form a regular sequence in* S *if and only if the algebraic variety*

$$V=V(f_1,\dots,f_p)=\{x\in\mathbb{C}^n;\ f_1(x)=\dots=f_p(x)=0\}$$

is a complete intersection (i.e. $\dim(V,x)=n-p$ *for any point* $x\in V$).

In particular, the property of such a sequence of being regular does not depend on the order of the polynomials in the sequence.

PROOF

We make this proof by induction on p. For $p=1$ everything is clear.

Assume from now on that the assertion above is true for $p-1$ and recall the notation S^i from the proof of (7.22).

Let $f_1,\dots,f_p$ be a regular sequence in S as above. Let $C_1\cup\dots\cup C_N$ be the decomposition of $V(f_1,\dots,f_{p-1})$ in irreducible components. Since the subsequence $f_1,\dots,f_{p-1}$ is regular, it follows that $\dim C_j=n-p+1$ for any j.

The fact that f_p is not a zero divisor in S^{p-1} is equivalent to $f_p|C_j\neq 0$ for any j. Indeed, let $q_1\cap\dots\cap q_N$ be the *minimal primary decomposition* of the ideal $(f_1,\dots,f_{p-1})$ in S [AM], such that $I(C_j)=\sqrt{q_j}$ for all j. If there is an element $a\in S$ such that $\hat{a}\neq 0$, $\hat{a}\cdot\hat{f}_p=0$ in S^{p-1}, then we must have $af_p\in q_j$ for any j. Since $\hat{a}\neq 0$, there is an index j_o such that $a\notin q_{j_o}$.

But then $f_p\in\sqrt{q_{j_o}}$ which is equivalent to $f_p|C_{j_o}=0$.

Coming back to our main proof, note that the irreducible components of the variety V are exactly the irreducible components of the various sets

$$C_{j,p}=\{x\in C_j;\ f_p(x)=0\}\quad \text{for } j=1,\dots,N$$

each of which has dimension $\dim C_j-1=n-p$. Therefore V is a complete intersection.

Conversely, let now V be a complete intersection and let again $C_1\cup\dots\cup C_N$ be the irreducible decomposition for $V(f_1,\dots,f_{p-1})$. Then $\dim C_j\geq n-p+1$ by general reasons. Assume that $\dim C_{j_o}>n-p+1$ for some j_o. Since the polynomials f_i are *weighted homogeneous*, it follows that all the components C_j are

$\mathbb{C}^*$-invariant (recall the proof of (7.8)!). In particular, $0 \in C_{j_o}$ and since $f_p(0)=0$ it follows that the set $C_{j_o,p}$ defined as above has dimension strictly greater than n-p. Since $C_{j_o,p} \subset V$, this is a contradiction. Hence $V(f_1,\ldots,f_{p-1})$ is also a complete intersection and by induction hypothesis we get that $f_1,\ldots,f_{p-1}$ is a regular sequence in S.

Moreover, it is clear that $f_p|C_j \neq 0$ for any j and hence, as in the first part of the proof, we infer that the whole sequence $f_1,\ldots,f_{p-1},f_p$ is regular in S.

(7.24) COROLLARY

A sequence $f_1,\ldots,f_n$ *of weighted homogeneous polynomials in* $\mathbb{C}[x_1,\ldots,x_n]$ *is regular if and only if* $V(f_1,\ldots,f_n)=\{0\}$.

(7.25) EXERCISE

A sequence of *real* polynomials $f_1,\ldots,f_p$ in $\mathbb{R}[x_1,\ldots,x_n] \subset \mathbb{C}[x_1,\ldots,x_n]$ is regular in $\mathbb{R}[x_1,\ldots,x_n]$ if and only if this sequence is regular in $\mathbb{C}[x_1,\ldots,x_n]$. Hence one can use (7.23) to decide if a given sequence of weighted homogeneous real polynomials is regular!

Assume from now on that $f \in H(\underline{w},d;K)$ is a finitely determined polynomial. Then to compute its Milnor number $\mu(f)$ we can assume $K=\mathbb{C}$ by (6.60). Then the partial derivatives $\frac{\partial f}{\partial x_1},\ldots,\frac{\partial f}{\partial x_n}$ form a regular sequence in $S=\mathbb{C}[x_1,\ldots,x_n]$ by (7.24) and (6.19.iv).

Moreover the Milnor algebra M(f) is naturally isomorphic to the quotient algebra $R=S/(\frac{\partial f}{\partial x_1},\ldots,\frac{\partial f}{\partial x_n})$. Indeed, the morphism

$$(7.26) \qquad j: R \longrightarrow M(f)$$

induced by the inclusion $S \subset E_n$ is well-defined. Next j is clearly surjective, since J_f is an ideal of finite codimension in E_n and hence one can find a monomial basis for the Milnor algebra M(f).

To prove that j is injective we use *the weighted homoge-*

neity of the function f. Assume that $g \in S$, with $j(\hat{g})=0$, $\hat{g} \in R$ being the corresponding class. This means that there is a relation in E_n

$$g=a_1\frac{\partial f}{\partial x_1}+\ldots+a_n\frac{\partial f}{\partial x_n}$$

for some elements $a_i \in E_n$. Taking the weighted homogeneous components with respect to $\underline{w}$ of this equality we deduce that any such component g_d of g belongs to the ideal generated by the partial derivatives in S. Hence $\hat{g}=0$ and j is an isomorphism.

Via this isomorphism of $\mathbb{C}$-algebras, the Milnor algebra M(f) becomes a graded $\mathbb{C}$-algebra and its Poincaré series (which is in fact a polynomial since M(f) is artinian) can be computed from (7.22). This gives us the next result.

(7.27) PROPOSITION

Let f *be a finitely determined weighted homogeneous polynomial of degree* d *with respect to the weights* $\underline{w}=(w_1,\ldots,w_n)$. *Then*

$$P_{M(f)}(t)=\frac{(1-t^{d-w_1})\ldots(1-t^{d-w_n})}{(1-t^{w_1})\ldots(1-t^{w_n})}$$

$$\mu(f)=P_{M(f)}(1)=\frac{(d-w_1)\ldots(d-w_n)}{w_1\ldots w_n}$$

In particular, both invariants are *independent* of the coefficients of f and hence we can use also the notations $P(\underline{w},d)$ for $P_{M(f)}$ and $\mu(\underline{w},d)$ for $\mu(f)$.

Then the condition on $P(\underline{w},d)$ to be a polynomial with positive integer coefficients is clearly a *necessary condition* for the existence of a finitely determined f in $H(\underline{w},d;K)$. But this condition is unfortunately *not sufficient* for $n\geq 4$ as follows from the next example, due to B.M. Iliev [AGV], p. 201. Take $n=4$, $w_1=1$, $w_2=24$, $w_3=33$, $w_4=58$, $d=265$. Then a direct computation of $P(\underline{w},d)$ shows that it satisfies the above condition, but a complete list of all the monomials of degree 265 with respect to the given weights implies that there is no finitely determined linear combination of them.

Another interesting property of the Poincaré polynomial $P(\underline{w},d)$ which follows directly from the first formula in (7.27) is that

(7.28) $$P(\underline{w},d)(t)=t^D P(\underline{w},d)(t^{-1})$$

where $D=D(\underline{w},d)=\deg P(\underline{w},d)=nd-2\sum_{i=1,n} w_i$. In other words, the coefficients $q_k=\dim M(f)_k$ of the polynomial $P(\underline{w},d)$ have the next *symmetry property*

(7.29) $$q_k=q_{D-k}\ ,\quad \text{for any}\quad k=0,1,\dots,D.$$

In particular, $q_D=q_o=1$ and it can be shown (see [AGV], p.102) that the 1-dimensional vector space $M(f)_D$ is spanned by the class of the *hessian* $H(f)$ of f defined as in (5.12).

Note that any monomial of degree strictly greater than D belongs to the Jacobian ideal J_f .

(7.30) EXAMPLE

Let f be a *homogeneous polynomial* in n-variables of degree $d\geq 2$. If f is finitely determined as a germ in E_n , then

$$P_{M(f)}(t)=\frac{(t^{d-1}-1)^n}{(t-1)^n}\ ,\ \mu(f)=(d-1)^n \quad \text{and} \quad D=n(d-2)\ .$$

The last equality shows that

$$\min\{s\in \mathbb{N}\ ;\ m_n^s\subset J_f\}=n(d-2)+1$$

for any degree $d\geq 2$.

(7.31) EXERCISE

Use the above to get the following *estimates for the orders of* $\mathcal{R}$ *and* $\mathcal{K}$ *determinacy of a homogeneous polynomial* f

$$o_{\mathcal{R}}(f)=o_{\mathcal{K}}(f)=\begin{cases} d & \text{for } d=2, \text{ or } n=1, \text{ or } (n,d)=(2,3) \\ n(d-2) & \text{otherwise} \end{cases}$$

Even more particularly, if $f=x_1^3+\dots+x_n^3$ for $n\geq 3$, one has $o_{\mathcal{R}}(f)=o_{\mathcal{K}}(f)=n$, which can be much greater than $\deg f=3$.

We end this section with a consequence of Proposition (7.27) and of Mather's Lemma.

Let $U(\underline{w},d)=H(\underline{w},d;\mathbb{C})\setminus A$ be the Zariski open set of the weighted homogeneous functions of type $(\underline{w},d)$ which are finitely determined.

(7.32) COROLLARY

Assume that $U(\underline{w},d)$ *is not empty. Then any two functions* $f,g\in U(\underline{w},d)$ *are R-equivalent if and only if* $q_d=0$, *where* q_d *is the coefficient of* t^d *in the polynomial* $P(\underline{w},d)(t)$.

PROOF

Note first that all the functions $f\in U(\underline{w},d)$ have the same Milnor number μ by (7.27). It follows by (6.19) and (2.10) that there is a positive integer k such that all the functions $f\in U(\underline{w},d)$ are k-R-determined. Moreover, we take k big enough such that any monomial $x^a=x_1^{a_1}\dots x_n^{a_n}$ in $H(\underline{w},d;\mathbb{C})$ satisfies $a_1+\dots+a_n\leq k$. Then there is a natural *linear inclusion* $H(\underline{w},d)\subset\subset J^k(n,1)$ and we are going to apply Mather's Lemma (4.26) in the situation: $G=R^k$, $M=J^k(n,1)$, $P=U(\underline{w},d)$. Note first that, since we are in the complex case, P is indeed connected. Next we get

$$T_{j^kf}(R^kj^kf)=j^k(mJ_f)=\frac{mJ_f+m^{k+1}}{m^{k+1}}=\frac{mJ_f}{m^{k+1}}$$

where the last equality comes from (6.3). It follows that

$$\operatorname{codim}(R^kj^kf)=\dim\frac{m}{m^{k+1}}-\dim\frac{mJ_f}{m^{k+1}}=\dim\frac{m}{mJ_f}=n+\mu-1 \quad \text{by (6.52 i).}$$

Hence all the orbits R^kj^kf, for $f\in U(\underline{w},d)$, have the same dimension.

Now if $U(\underline{w},d)\subset R^kj^kf$ for some $f\in U(\underline{w},d)$ it follows, passing to tangent spaces, that

$$\frac{H(\underline{w},d)+m^{k+1}}{m^{k+1}}\subset\frac{mJ_f}{m^{k+1}}$$

This clearly implies $H(\underline{w},d)\subset J_f$, in other words $q_d=0$, since all the monomials of degree d are in J_f. Conversely, if $H(\underline{w},d)\subset J_f$ we get in fact $H(\underline{w},d)\subset mJ_f$ (note that the partial derivatives of f are weighted homogeneous of degrees $d-w_1,\dots,d-w_n$). This shows as above that $T_{j^kf}U(\underline{w},d)\subset T_{j^kf}(R^k.j^kf)$ for all $f\in U(\underline{w},d)$. Then we can apply Mather's Lemma and deduce that $j^kf\sim j^kg$ for any $f,g\in U(\underline{w},d)$. By the choice of k, this shows that f and g are $\mathcal{R}$-equivalent.

(7.33) EXERCISE

Show by an example that one can have $U(\underline{w},d)\neq\emptyset$ and $q_d=0$ even if $d<D$, with D as in (7.28). Hint: it is enough to look at the case n=2.

§3. SEMI WEIGHTED HOMOGENEOUS SINGULARITIES

In this section we describe a larger class of singularities than the weighted homogeneous ones, but having many similar properties. We are going to use the notations from the previous sections.

(7.34) DEFINITION

A map germ $g\in E^0_{n,p}$ is called *semi weighted homogeneous* (or *semi quasi homogeneous*) of type $(\underline{w},\underline{d})$ if there is a finitely $\mathcal{K}$-determined weighted homogeneous polynomial map $f\in U(\underline{w},\underline{d};\mathcal{K})$ such that any monomial which occurs in $h_i=g_i-f_i$ has degree (with respect to $\underline{w}$) strictly greater than d_i.

Here $\underline{w}=(w_1,\dots,w_n)$, $\underline{d}=(d_1,\dots,d_p)$ and f_i,g_i and h_i denote the i-th component of f,g and h=g-f respectively. (In the real smooth case one should look at the monomials which occur in the Taylor expansion of h_i at the origin!)

The set of all semi weighted homogeneous map germs associated in the above way to a given $f\in U(\underline{w},\underline{d};\mathcal{K})$ will be denoted by $U^+(\underline{w},\underline{d};\mathcal{K})_f$. And the set of all semi weighted homogeneous map germs of type $(\underline{w},\underline{d})$, denoted by $U^+(\underline{w},\underline{d};\mathcal{K})$, is precisely the union $\bigcup U^+(\underline{w},\underline{d};K)_f$, when $f \in U(\underline{w},\underline{d};K)$. Here $U(\underline{w},\underline{d};K) \subset \subset H(\underline{w},\underline{d};K)$ is the Zariski open set of finitely K-determined polynomial mappings.

(7.35) PROPOSITION

If $g \in U^+(\underline{w},\underline{d};K)_f$, then g is finitely K-determined.

PROOF

For $t \in K^*$ we consider the diffeomorphism germ $h_t \in D_n$, with $h_t(x) = (t^{-1}) \cdot x$ as in (7.4) and the matrix $A_t = (a_{ij}) \in \in M_{n,p}$ with $a_{ij} = 0$ for $i \neq j$ and $a_{ii} = t^{-d_i}$ for $i = 1, \dots, p$.

Then it follows that $g = f + h$ is K-equivalent to the germ

$$g_t(x) = A_t \cdot (g \circ h_t^{-1}) = f(x) + h(x,t) \tag{7.36}$$

where $\lim_{t \to 0} h(x,t) = 0$. Indeed any monomial in $h(x,t)$ contains some positive power of t by (7.34). It follows that in any jet space $J^k(n,p)$ one has the specialization $j^k g \to j^k f$. Hence $\operatorname{codim}(K^k_{n,p} \cdot j^k g) \leq \operatorname{codim}(K^k_{n,p} \cdot j^k f)$ which by (6.39) implies that g is finitely K-determined.

In the case of function germs one has the next two basic results (a proof for the first can be found in [Lm] p.199 and for both in [AGV], p.198 and p. 209). Moreover, the reader can easily prove (7.38), by using the idea of the proof of Proposition (7.41) below.

(7.37) PROPOSITION

For a function germ $g \in U^+(\underline{w},d;K)_f$ one has $\mu(g) = \mu(f)$.

(7.38) PROPOSITION

Assume that $f \in U(\underline{w},d;K)$ and that $u_1, \dots, u_s$ are monomials in S of degree strictly greater than d and such that their classes in the Milnor algebra $M(f)$ form a basis for the subspace $\bigoplus_{k>d} M(f)_k$. Then any function germ $g \in U^+(\underline{w},d;K)_f$ is R-equivalent to a polynomial of the form

$$f + c_1 u_1 + \dots + c_s u_s \quad \textit{for some} \quad c_i \in K.$$

In particular, if $M(f)_k=0$ *for all* $k>d$ *(equivalently* $D(\underline{w},d)\leq d$*), then any function germ in* $U^+(\underline{w},d;K)_f$ *is* R*-equivalent to* f.

(7.39) COROLLARY

Assume that $U(\underline{w},d;\mathbb{C})$ *is not empty. Then the function germs in* $U^+(\underline{w},d;\mathbb{C})$ *are contained in a single* R*-orbit if and only if* $D(\underline{w},d)<d$.

PROOF

If $D(\underline{w},d)<d$, then the result follows by (7.32) and (7.38).

Assume now that $D=D(\underline{w},d)\geq d$ and let $u\in S_D$ such that $u\notin J_f$, for some $f\in U(\underline{w},d;\mathbb{C})$. Then the function $f_t=f+tu$ is clearly in $U^+(\underline{w},d;\mathbb{C})$ for t small enough. If this set is contained in an R-orbit, we get as above that

$$u=\frac{df_t}{dt}\in TRf=mJ_f$$

a contradiction.

(7.40) REMARK

The analogue of (7.37) for *Tjurina numbers* is false in general. For example, let $f\in U(\underline{w},d;K)$ and $h\in S_e$ with $e>d$ and such that $h\notin J_f$. Then consider the family of germs g_t as in (7.36). Note that $g_t\in U^+(\underline{w},d;K)$ for any $t\in K^*$ and $g_o=f$. Hence the equality $\tau(g_t)=\tau(f)$ would imply the equality of orbits $K^s_{n,1}j^sg_t=K^s_{n,1}j^sf$ for any large integer s.

If we put $\overline{g}_t=f+th$, the equality of orbits above implies

$$h=\left.\frac{d\overline{g}_t}{dt}\right|_{t=0}\in j^s(TKf)=\frac{mJ_f+m^{s+1}}{m^{s+1}}$$

For s large enough, $mJ_f\supset m^{s+1}$ and we get $h\in J_f$, a contradiction.

On the other hand, we remark that there is an analogue of Proposition (7.38) for the K-equivalence of map germs in $U^+(\underline{w},\underline{d};K)_f$. To present it, we need first some notations.

Let k be a sufficiently large positive integer such that one has a linear inclusion $H(\underline{w},\underline{d};K)\subset J^k(n,p)=J^k$.

A *monomial* in J^k is by definition an element of the form $u=a\cdot e_j$ where $a\in S_e$ is a monomial, for some $e=\deg(a)$, and e_j is an element of the standard basis: $e_j=(0,\dots,0,1,0,\dots,0)$ with 1 on the position j.

We put the *degree* of such a monomial u to be $\deg u=\deg(a)-d_j=e-d_j\in\mathbb{Z}$.

Then there is a *grading* of J^k associated to the weighted homogeneity type $(\underline{w},\underline{d})$, namely

$$J^k=\bigoplus_{s\in\mathbb{Z}} J^k_s$$

where J^k_s is the vector subspace spanned by all the monomials u of degree s. Note that $H(\underline{w},\underline{d};K)=J^k_o$ and put $J^k_>=\bigoplus_{s>0} J^k_s$. For a weighted homogeneous polynomial map $f\in J^k_o$ the corresponding contact tangent space $T^k=j^k(TKf)$ is a homogeneous vector subspace in J^k [i.e. $u=\sum u_s$ (with $u_s\in J^k_s$) is in T^k if and only if $u_s\in T^k$ for all s].

It follows that the quotient vector space $N^k(f)=J^k/T^k$ has an induced *grading*.

The notation $N^k(f)_>=\bigoplus_{s>0} N^k(f)_s$ is used in the next statement.

(7.41) PROPOSITION

With the above notations, let $u_1,\dots,u_s$ *be a set of monomials in* $J^k_>$ *whose classes in* $N^k(f)_>$ *form a basis. Any* k-jet $g=f+v$ *with* $v\in J^k_>$ *is* $K^k_{n,p}$*-equivalent to a* k*-jet of the form*

$$\bar{f}=f+c_1u_1+\dots+c_su_s \quad \textit{for some} \quad c_i\in K.$$

In particular, if $f\in U(\underline{w},\underline{d};K)$ *then any map germ* $g\in U^+(\underline{w},\underline{d};K)_f$ *is K-equivalent to a polynomial map germ of the form* $\bar{f}$.

PROOF

We can assume that $d_1\geq d_2\geq\dots\geq d_p$. Let $v=v_o+v_1$, where $v_o\in J^k_e$, $v_1\in\bigoplus_{s>e} J^k_s$ for some integer $e>0$. Let u_j for $j\in I$ be the elements of degree e in the given set of monomials. Then, for some constants $c_j\in K$, one has

$$\overline{v}_o = v_o - \sum_{j\in I} c_j u_j \in T_e^k$$

This implies that there are some elements $a_i \in S_{e+w_i}$ and $b_{ij} \in S_{e+d_i-d_j}$ such that

$$\overline{v}_o = \sum_{i=1,n} a_i \frac{\partial f}{\partial x_i} + B\cdot f$$

where B is the p×p matrix (b_{ij}).

Consider the isomorphism germ $h\in D_n$ given by $h(x) = (x_1+a_1(x),\dots,x_n+a_n(x)) = x+a(x)$. Note that its inverse satisfies the following $h^{-1}(x) = x-a(x)+b(x)$, where all the monomials in the component $b_i(x)$ have degree $> e+w_i$.

We also consider the matrix $A = 1_p - B$ in $M_{n,p}$ (indeed A is invertible since $\deg b_{ij} > 0$ for $i \geq j$ and hence $A(0)$ is a triangle matrix with $\det A(0) = 1$).

Then an easy computation using Taylor's formula shows that

$$g' = (h,A)\cdot g = A(g\circ h^{-1}) = f + v_o - \sum_{i=1,n} a_i \frac{\partial f}{\partial x_i} - B\cdot f + \overline{v}_1$$

where $\overline{v}_1 \in \bigoplus_{s>e} J_s^k$.

In other words: $g \sim g' = f + \sum_{j\in I} c_j u_j + \overline{v}_1$.

We can obviously iterate this step (i.e. replace v by $\overline{v}_1$) and in a finite number of such steps we get the complete proof of (7.41).

We mention in the end of this section that *semi weighted homogeneous singularities occur naturally in the classification of unimodal singularities of functions* ([AGV], p.246) *and of mappings* [DGn], n=2,3,4.

§4. AROUND A THEOREM OF K. SAITO

The important role played by weighted homogeneous singularities goes far beyond the possibility of making some nice explicit computations as in (7.27).

In fact, in *Singularity Theory* one is not concerned with a single function germ, but usually with a whole R or K equivalence class. Hence it is a natural question to ask for a given function germ f whether or not there is a weighted homogeneous polynomial f_o (with respect to some a priori unknown weights) which is R or K equivalent to f.

In the affirmative case, it is natural to take f_o as a *normal form* (representative) for the orbit Rf or Kf of the germ f.

The answer to this important question is due to K. Saito [S1] and we incorporate it in the next long statement.

(7.42) THEOREM

For a complex analytic function germ $f \in m \subset E_n$ *which is finitely determined, the following statements are equivalent.*

(i) f *is R-equivalent to a weighted homogeneous polynomial* f_o.

(ii) f *is K-equivalent to a weighted homogeneous polynomial* f_o.

(iii) $Rf=Kf$.

(iv) *For any* $s \geq 1$, $\operatorname{codim} R^s \cdot j^s f = \operatorname{codim} K^s_{n,1} \cdot j^s f$ *where both codimensions are with respect to* $J^s(n,1)$. *In other words, the foliation of a sufficient K-orbit for* f *by the corresponding R-orbits (see* (6.55)) *is trivial* (i.e. *contains just one leaf*).

(v) $TRf=TKf$

(vi) $f \in mJ_f$

(vii) $f \in J_f$

(viii) $\mu(f)=\tau(f)$.

PROOF

The implications (i)$\Longrightarrow$(ii), (vii)$\Longleftrightarrow$(viii) and (iii)$\Longrightarrow$ $\Longrightarrow$(iv)$\Longleftrightarrow$(v)$\Longleftrightarrow$(vi)$\Longrightarrow$(vii) are just direct consequences of the definitions.

In what follows, we prove *all the other implications, with the exception of the single really difficult one*, namely (vii)$\Longrightarrow$(i) which is the object of Saito's paper [S1]. For a more precise statement for this implication, see (7.43) below.

STEP 1 (ii)$\Longrightarrow$(i)

Assume that $f=a\cdot(f_o \circ h)$, where $f_o \in \mathbb{C}[x_1,\dots,x_n]$ is weighted homogeneous of degree d with respect to some weights $\underline{w}=$ $=(w_1,\dots,w_n)$, $a \in E_n$ is an invertible element and $h \in D_n$ is an isomorphism germ. Since $a(0)\neq 0$, there is a germ $b \in E_n$ such that $b^d=a$. In fact, this equation has d different solutions obtained from a fixed one by multiplication with the d-roots of the unity.

Since $b(0)\neq 0$, the map germ

$$\bar{h}:x \longmapsto b(x)\cdot h(x)=(b(x)^{w_1}\cdot h_1(x),\dots,b(x)^{w_n}\cdot h_n(x))$$

is an element in D_n, where h_i denote the components of the germ h.

Then one obviously has $f=f_o \circ \bar{h}$, showing that f is $\mathcal{R}$-equivalent to f_o.

STEP 2

(i) and (ii)$\Longrightarrow$(iii).

By (i) and (ii) we have $\mathcal{R}f=\mathcal{R}f_o$, $\mathcal{K}f=\mathcal{K}f_o$. Hence what we must prove is $\mathcal{R}f_o=\mathcal{K}f_o$. But the nontrivial inclusion $\mathcal{K}f_o \subset \mathcal{R}f_o$ was just proved in Step 1.

STEP 3

(iv)$\Longrightarrow$(iii).

Note that a function germ f is s-$\mathcal{R}$-determined if and only if $\mathcal{R}f=(j^s)^{-1}(R^s\cdot j^sf)$, where $j^s:E_n^o \to J^s(n,1)$ denotes as usual the natural projection. And a similar statement holds for $\mathcal{K}$-equivalence.

Therefore it is enough to show that $R^s\cdot j^sf=K^s_{n,1}\cdot j^sf$, at least for s large enough. But this is true for any $s\geq 1$, using the following simple remarks

(a) $R^s\cdot j^sf \subset K^s_{n,1}\cdot j^sf$.

(b) $\dim(R^s\cdot j^sf)=\dim(K^s_{n,1}\cdot j^sf)$ by (iv).

(c) Any R^s or $K^s_{n,p}$ orbit is *connected in the complex case.* Indeed these groups are connected since $G\ell(n,\mathbb{C})$ is connected and one can proceed by induction on s. To do this, note that the kernels of the natural projections

$$R^s \to R^{s-1} \quad \text{and} \quad K^s_{n,p} \to K^{s-1}_{n,p}$$

are all, in a natural way, complex vector spaces and hence connected (Exercise for the reader!).

STEP 4

(vii)$\Longrightarrow$(vi).

It is enough to show that a relation

$$f=a_1 \frac{\partial f}{\partial x_1}+\ldots+a_n \frac{\partial f}{\partial x_n}$$

where the coefficients $a_i \in E_n$ are not all in m leads to a contradiction. For this, see once again the proof of (6.52.ii).

Assuming that (vii)$\Longrightarrow$(i), this ends the proof of the Theorem.

It is perhaps worthwhile to mention explicitly that not every function germ is equivalent to a weighted homogeneous polynomial. For instance, the function $f=x^2y^2+x^5+y^5$ considered in Example (6.56) has $\mu(f)>\tau(f)$ and hence, by the above Theorem, f is not equivalent to a weighted homogeneous polynomial. See also (7.40).

On the other hand, some polynomials can be weighted homogeneous in *several distinct ways*. The simplest example is $g=x_1x_2+x_3^2$ which is weighted homogeneous of degree 2k for any weights $w=(w_1,w_2,w_3)$ with $w_1+w_2=2k$ and $w_3=k$.

Assume from now on in this Chapter that the weights $\underline{w}=(w_1,\ldots,w_n)$ which occur satisfy all the condition

$$G.C.D.(w_1,\ldots,w_n)=1 \quad \text{as in (7.2).}$$

For the above polynomial g, the system of weights $\underline{w}=(1,5,3)$ and $\underline{w}'=(2,4,3)$ show that this additional condition is not enough to guarantee the uniqueness.

The way to avoid this ambiguity is clarified in the same paper by K. Saito [S1].

(7.43) THEOREM

Assume that $f \in m^2$ *is a finitely determined complex function germ satisfying the condition* $f \in J_f$. *Then* f *is R-equivalent to a weighted homogeneous polynomial* $f_o \in \mathbb{C}[x_1,\ldots,x_n]$ *of type* $(\underline{w},d)$ *such that*

$$f_o(x)=h(x_1,\ldots,x_r)+x_{r+1}^2+\ldots+x_n^2$$

where h *is a weighted homogeneous polynomial in* $\mathbb{C}[x_1,\ldots,x_r]$ *of degree* d *with respect to the weights* $(w_1,\ldots,w_r)$, $j^2h=0$, $2w_i<d$ *for all* $i=1,\ldots,r$ *and* $2w_j=d$ *for* $j=r+1,\ldots,n$.

Moreover the positive integer r, $0\le r\le n$, *and the weighted homogeneity type* $(\underline{w},d)$ *are uniquely determined by the germ* f *and the conditions above.*

In fact, the integer r is called the *corank* of the function germ f and can be defined for any germ $f \in m_n^2$ by the formula

$$\text{(7.44)} \qquad r=\text{corank}(f)=n-\text{rank}\left(\frac{\partial^2 f}{\partial x_i \partial x_j}(0)\right)_{i,j=1,\ldots,n} .$$

(7.45) COROLLARY

Let f_o *be a finitely determined weighted homogeneous polynomial as in* (7.43). *Then its weighted homogeneity type satisfies the condition* $D(\underline{w},d)<d$ *if and only if* h *is equivalent to one of the normal forms in the next Table.*

Symbol	corank (f_o)	h	Condition
A_1	0	0	
A_k	1	x_1^{k+1}	$k\ge 2$
D_k	2	$x_1^2x_2+x_2^{k-1}$	$k\ge 4$
E_6	2	$x_1^3+x_2^4$	
E_7	2	$x_1^3+x_1x_2^3$	
E_8	2	$x_1^3+x_2^5$	

Note that the subscripts indicate the corresponding Milnor numbers $\mu(h)=\mu(f_o)$. The use of the symbols A, D and E will be explained latter in this book.

PROOF

The cases r=corank $(f_o)=0$ or 1 are obviously clear.

Then the case r=2 follows by a case by case analysis of the polynomials described in Table (7.19). Finally we have to show that the case $r\geq 3$ cannot occur. The inequality $D(\underline{w},d)<d$ is clearly equivalent to

(7.46) $$\left(\frac{r-1}{2}\right)d<w_1+w_2+\ldots+w_r$$

We show now that (7.46) is impossible for r=3 and refer the reader for the complete proof to [S2], (Satz 2.11).

Let r=3 and assume that $w_1\geq w_2\geq w_3$. Then by the condition of finite determinacy, one cannot have $f\in(x_2,x_3)^2$, the square of the ideal in E_3 generated by x_2 and x_3. It follows that f must contain a monomial of the form x_1^k or $x_1^{k-1}x_i$ for $k\geq 3$ and i=2 or 3.

But then $d=\deg(f)=kw_1\geq w_1+w_2+w_3$ or $d=(k-1)w_1+w_i\geq w_1+w_2+w_3$. Hence in both cases, the inequality (7.46) is impossible.

The importance of this Corollary is related to classifying semi weighted homogeneous function germs (recall (7.39)) and will become even more clear in the next Chapter.

§5. UNIQUENESS OF WEIGHTED HOMOGENEITY TYPE

We have seen in the end of the previous section that for a finitely determined weighted homogeneous function germ its weighted homogeneity type $(\underline{w},d)$ is unique if we normalize it properly. In this section we extend this type of result to the case of mappings, using a deep result due to C.T.C. Wall [W1].

Let $f\in U(\underline{w},\underline{d};\mathbb{C})$, where $n>p$, $\underline{w}=(w_1,\ldots,w_n)$, $\underline{d}=(d_1,\ldots,d_p)$ be a mapping such that $j^1f=0$. This means that the *isolated com-*

plete intersection singularity $X=f^{-1}(0)$ *is minimally embedded in* $\mathbb{C}^n$, i.e. $I_X=I_f\subset m^2$. In these conditions we say that $(X,0)$ is *weighted homogeneous of type* $(\underline{w},\underline{d})$.

Note that n=edim X and p=edim X-dim X depend clearly only on the isomorphism class of $(X,0)$ (3.11.ii). What is not clear at all is that even the weighted homogeneity type $(\underline{w},\underline{d})$ is determined by this isomorphism class.

More precisely, let $X'=(f')^{-1}(0)$ be another isolated complete intersection singularity, for $f'\in U(\underline{w}',\underline{d}';\mathbb{C})$, with $\underline{w}'=(w_1',\dots,w_n')$, $\underline{d}'=(d_1',\dots,d_p')$. With these notations, we have the next result.

(7.47) THEOREM

Assume that the singularities $(X,0)$ *and* $(X',0)$ *are isomorphic. If either*

(i) $p=\operatorname{codim}(X)=\operatorname{codim}(X')>1$, *or*

(ii) $p=1$ *and* $\operatorname{mult}(X)\geq 3$

then the weighted homogeneity types $(\underline{w},\underline{d})$ *and* $(\underline{w}',\underline{d}')$ *coincide (up to permutations of weights and degrees) and there is a* $\mathbb{C}^*$*-equivariant isomorphism* $h:(X,0)\to(X',0)$.

Before starting the proof, we make some comments on this statement.

It follows from (4.5) that an isolated singularity of strictly positive dimension is reduced. Hence our germs $(X,0)$ and $(X',0)$ are both reduced and we can use any notion of isomorphism from (3.16).

To explain the notation from (ii), we define for a hypersurface singularity $(X,0)$ its *multiplicity* mult(X) in the following way. Take f a generator for the principal ideal $I_X\subset E_n$ and put

(7.48) $$\operatorname{mult}(X)=\operatorname{ord}(f)=\min\{k;\ j^kf\neq 0\}$$

We leave as an *exercise* for the reader to show that this definition is correct (i.e.does not depend on the choice of the generator f!)

The final assertion about h means that this map germ has a polynomial extension $\bar{h}:(\mathbb{C}^n,0)\to(\mathbb{C}^n,0)$ such that $\bar{h}(t\cdot x)=t\cdot\bar{h}(x)$, where the points stand for the $\mathbb{C}^*$-action introduced in (7.4) (i.e. the i-component $\bar{h}_i$ of $\bar{h}$ is a weighted homogeneous polynomial with respect to $\underline{w}$ of degree w_i, for all $i=1,\dots,n$.)

PROOF

For a map germ $g \in E^o_{n,p}$ we consider its K-isotropy subgroup, namely

$$K_g=\{(a,A)\in K_{n,p};\ (a,A)\cdot g=g\}$$

Since $(X,0)$ and $(X',0)$ are isomorphic, it follows that f and f' are in the same K-orbit by (3.16) and hence their isotropy subgroups K_f and $K_{f'}$ are conjugated inside $K_{n,p}$ via some element $(\bar{a},\bar{A})\in K_{n,p}$.

Since f is weighted homogeneous of type $(\underline{w},\underline{d})$, there is a group monomorphism $q:\mathbb{C}^*\to K_f$, $q(t)=(a_t,A_t)$ where $a_t(x)=t^{-1}\cdot x$ as in (7.4) and A_t is the diagonal matrix $\mathrm{diag}(t^{-d_1},\dots,t^{-d_p})$ (compare to the proof of (7.35)).

Similarly we get a monomorphism $q':\mathbb{C}^*\to K_{f'}$ and up to conjugation by $(\bar{a},\bar{A})$, we consider $q':\mathbb{C}^*\to K_f$. Then by *Theorem* (3.3) in [W1] we get that q and q' are conjugated by some element $(b,B)\in K_f$. If we write everything explicitly , this means that

$$h=\bar{a}\circ b:(X,0)\xrightarrow{\sim}(X,0)\xrightarrow{\sim}(X',0)$$

is a $\mathbb{C}^*$-invariant isomorphism.

Next, this isomorphism h induces an isomorphism of linear $\mathbb{C}^*$-actions

$$\mathbb{C}^n=m_{X'}/m^2_{X'}\xrightarrow{h^*} m_X/m^2_X=\mathbb{C}^n$$

where $m_X\subset \mathcal{O}_X$, $m_{X'}\subset \mathcal{O}_{X'}$ denote the maximal ideals. The first $\mathbb{C}^*$-action above corresponds to the weights $\underline{w}'$ and the second to $\underline{w}$. This shows that $\underline{w}'=\underline{w}$, up to the order. Since $h^*(I_{X'})=I_X$, it follows easily that $\underline{d}=\underline{d}'$ and hence we are ready.

(7.49) EXAMPLE

Let us consider the *family of plane curve singularities* $X_t:f_t=x^3+x^2y^m+ty^{3m}$, where $m\geqslant 2$, $t\neq 0$, $-\frac{4}{27}$. This family is denoted by the symbol $E_{m,0}$ in Wall's list from [W3]. Note that for $t=0$ or $t=-\frac{4}{27}$ the germ $(X_t,0)$ has a double branch (i.e. it is not reduced) and hence does not define an isolated singularity.

We want to decide for which values $s,t\in\mathbb{C}\setminus\{0,-\frac{4}{27}\}$, one has an isomorphism $X_s\sim X_t$. If we put $\underline{w}=(w_1,w_2)$, $w_1=wt(x)=m$, $w_2=wt(y)=1$, $d=3m$, then all the singularities X_t are weighted homogeneous of type $(\underline{w},d)$. Since we also have $\text{mult}(X_t)=3$, we can apply the above result.

It follows that if $X_s\sim X_t$, then there is a polynomial isomorphism $h:\mathbb{C}^2\to\mathbb{C}^2$ such that $\deg h_1=w_1$, $\deg h_2=w_2$ and such that $kf_s=f_t\circ h$, for some $k\in\mathbb{C}^*$.

The condition on the components h_i of h implies that

$$h_1(x,y)=bx+cy^m,\ h_2(x,y)=ay,\ \text{for } a,b,c\in\mathbb{C},\ ab\neq 0.$$

The second condition is equivalent to the system of equations:

$$b^3=k,\ 3b^2c+a^mb^2=k,\ 3bc^2+2a^mbc=0\ \text{and}\ c^3+c^2a^m+ta^{3m}=ks$$

Assume first that $c=0$. Then $a^m=b=k^{1/3}$ from the first two equations and $t=s$ using the last one.

Now let $c\neq 0$. Then the third equation gives $c=-2a^m/3$ and next the second yields $a^m=-k^{1/3}$. If we substitute all this in the last equation we get $t+s=-\frac{4}{27}$. This shows that the rational function

$$j:\mathbb{C}\setminus\{0,-\tfrac{4}{27}\}\to\mathbb{C}\setminus\{0\},\ j(t)=t^2+\frac{4t}{27}$$

is a *basic invariant* for the family X_t of singularities (i.e. $X_s\sim X_t \Longleftrightarrow j(s)=j(t)$).

CHAPTER 8

CLASSIFICATION OF SIMPLE SINGULARITIES OF FUNCTIONS

§1. SIMPLE SINGULARITIES AND ADJACENCY

By the *classification of the singularities* belonging to some subset $\Sigma \subset E^0_{n,p}$ of finitely determined map germs f one understands:

(i) the determination of the *normal forms* i.e. of some polynomial representatives as simple as possible [e.g. (semi) weighted homogeneous or involving only few monomials] for the equivalence classes $\mathcal{R}f$ or $\mathcal{K}f$ when $f \in \Sigma$.

(ii) the description of some *partial order* (*hierarchy*) on the set of these equivalence classes which is related to the adjacency of orbits defined in (5.1) and which says roughly which singularity is more complicated than another.

The subset $\Sigma \subset E^0_{n,p}$ above can be defined either by using some numerical invariants of the singularities, e.g.

$$\Sigma_k = \{f \in m_n \ ; \ \mu(f) \leq k\}$$

or by some more subtle properties, as in the next definition, due to Arnold (compare to [AGV], p.184).

(8.1) DEFINITION

A finitely $*$-determined map germ $f \in E^0_{n,p}$ is called *$*$-simple* if for any integer $s \geq 1$ the jet $j^s f$ has a neighbourhood in $J^s(n,p)$ which intersects only finitely many $*$-orbits. Here $* = \mathcal{R}, \mathcal{K}$.

It is natural to attempt first the classification of the simple singularities since such a classification, by the above definition, is discrete, i.e. does not involve parameters or moduli.

(8.2) EXERCISE

Show that for $p>1$, the only R-simple map germs in $E^o_{n,p}$ are the submersions. Hint: Consider the action r^1_n from (3.20) on $J^1(n,p)=\text{Hom}(K^n,K^p)$ and show that the class of a linear map $u:K^n \to K^p$ is uniquely determined by its image $u(K^n)\subset K^p$.

In this Chapter we classify the R and K simple function singularities. This result is due to Arnold [AGV], p. 184 and leads to the famous A - D- E singularities. These singularities play a central role in many parts of Singularity Theory and Algebraic Geometry and were studied by many outstanding mathematicians (we refer to the nice survey by Durfee [Du] and to the book of Lamotke [Lm] for more details). In our book, the reader will find additional information on curve and surface A - D -E singularities in Chapter 10.

Now, in this first section, we just present some preliminary results and introduce the adjacency of singularities. We work here only with K-equivalence, but the reader will find no difficulty in translating everything for R-equivalence.

For a finitely K-determined map germ $f\in E^o_{n,p}$ we introduce the next analogue of Tjurina number

$$\text{(8.3)} \qquad \text{codim}\ (f)=\dim\ (E^o_{n,p}/TKf)\ .$$

Then, for s large enough, the codimension of the $K^s_{n,p}$-orbit of j^sf in $J^s(n,p)$ is precisely equal to codim (f).

Using an analogue of (2.10) and (6.27), it follows that $s \geqslant \text{codim}\ (f) \Longrightarrow TKf\supset m^sE^o_{n,p} \Longrightarrow f$ is strongly $(s+1)$-K-determined $\Longrightarrow$ f is $(s+1)$-K-determined.

This remark leads to the next result

(8.4) LEMMA

Let $f\in E^o_{n,p}$ *be finitely* K*-determined and* $s\geq \text{codim}(f)+1$. *Then there is a Zariski neighbourhood* U *of* j^sf *in* $J^s(n,p)$ *such that any jet* $u\in U$ *is* K*-sufficient.*

PROOF

Consider the semialgebraic set

$$U=\{u\in J^s(n,p);\ \operatorname{codim}(K^s_{n,p}u)\leq \operatorname{codim}(f)\}.$$

To show that U is a Zariski neighbourhood of j^sf it is enough to show that U is open. And this follows from a *general fact* concerning the smooth action of a Lie group G on a smooth manifold M: any point $x\in M$ has a neighbourhood V in M such that $\dim(G\cdot y)\geqslant\dim(G\cdot x)$ for all $y\in V$. This in turn follows from (4.19) which implies that the minimal slice S intersects *transversally* all the orbits in a neighbourhood V of x.

Let $u=j^sg\in U$ and consider the sequence

$$E^o_{n,p}\supset TKg+mE^o_{n,p}\supset\ldots\supset TKg+m^sE^o_{n,p}$$

If all these inclusions are strict, it follows that

$$\operatorname{codim}(K^s_{n,p}\cdot u)=\dim\frac{E^o_{n,p}}{TKg+m^sE^o_{n,p}}\geq s$$

which is impossible by the definition of U.

Then for some $k\leq s$ one has equality

$$TKg+m^{k-1}E^o_{n,p}=TKg+m^kE^o_{n,p}$$

which implies $TKg\supset m^{k-1}E^o_{n,p}$ and, as above, g is k-determined. In particular, the jet u is K-sufficient and this ends the proof.

(8.5) DEFINITION

A neighbourhood U as in (8.4) is called a *K-sufficient neighbourhood* for the map germ f (or for its s-jet j^sf).

A first motivation for introducing this notion is the next.

(8.6) EXERCISE

Show that a finitely K-determined map germ $f\in E^o_{n,p}$ is K-simple if and only if there is a K-sufficient neighbourhood for f, in some jet space $J^s(n,p)$, which intersects finitely many $K^s_{n,p}$-orbits.

The second application of sufficient neighbourhoods is in relation with the following fundamental concept, which was introduced in a finite dimensional setting in (5.1).

(8.7) DEFINITION

We say that *an orbit* Kf (resp. *the singularity, normal form* $f \in E^{o}_{n,p}$) *specializes to an orbit* Kg (resp. *to the singularity, normal form* $g \in E^{o}_{n,p}$) if $\overline{K^{s}_{n,p} \cdot j^{s}f} \supset K^{s}_{n,p} \cdot j^{s}g$ for all $s \geq 1$. This situation is also described by saying that *the orbit* Kg (*or the singularity* g) *is adjacent to the orbit* Kf (*or to the singularity* f).

Moreover, we usually assume in this Definition that the orbits Kf and Kg are *distinct* and then use the notations $f \to g$, $Kf \to Kg$ to denote specializations.

Note that the relation of specialization is *transitive* (i.e. $f_1 \to f_2$ and $f_2 \to f_3$ implies $f_1 \to f_3$). In a certain intuitive way, $f \to g$ means that the singularity f is somehow "embedded" in the more complicated singularity g (and maybe this is a reason for the use of the symbol $f \to g$!). This will be made more precise in Example (8.16) below.

(8.8) EXERCISE

Let $g \in E^{o}_{n,p}$ be finitely K-determined and U a K-sufficient neighbourhood for g in some jet space $J^{s}(n,p)$. Then, for a map germ $f \in E^{o}_{n,p}$, the specialization $f \to g$ holds if and only if

$$j^{s}g \in \overline{(K^{s}_{n,p} \cdot j^{s}f \cap U)} \setminus (K^{s}_{n,p} \cdot j^{s}f) .$$

The next result is the first quantitative justification of our claim that $f \to g$ implies the singularity g is more complicated than the singularity f.

(8.9) LEMMA

(i) $Kf \to Kg$ *for map germs* $f,g \in E^{o}_{n,p}$ *implies*

$\operatorname{codim}(f) < \operatorname{codim}(g)$.

(ii) $Kf \to Kg$ *for function germs* $f,g \in m_n$ *implies*

$\mu(f) \leq \mu(g)$ *and* $\tau(f) < \tau(g)$.

(iii) *$Rf \to Rg$ for function germs $f,g \in m_n$ implies $Kf \to Kg$, $\mu(f)<\mu(g)$ and $\tau(f)<\tau(g)$.*

PROOF

(i) Follows directly from (8.8) and (4.9), once we remark that codim $(f)=\text{codim}\,(K^s_{n,p}\cdot j^s f)$ and similarly for g.

(ii) For a finitely determined function germ $f \in m_n$ one has

$$\tau(f)=\dim\,(E_n/TKf)-n=\text{codim}\,(f)+1-n$$

by (6.52 ii)).

Then the assertion for the Tjurina numbers is clear by (i).

To prove the assertion for the Milnor numbers, assume that $j^s g \in \overline{K^s_{n,1}\cdot j^s f} \cap U$ in the notations from (8.8). Then, for any small ball B centered at the origin $0 \in J^s(n,1)$, we can find a jet $j^s u \in B$ such that $j^s(g+u) \in K^s_{n,1}\cdot j^s f \cap U$.

Moreover, one can take U to be in the same time an R-sufficient neighbourhood for g. Then one has by (6.51) the equalities

$$\mu(f)=\mu(g+u)=\dim \frac{E_n}{J_{g+u}+m^{s+1}} \quad \text{and}$$

$$\mu(g)=\dim \frac{E_n}{J_g+m^{s+1}}$$

And a Linear Algebra argument in the finite dimensional vector space E_n/m^{s+1} shows that

$$\mu(f)=\lim_{u\to 0} \mu(g+u)\leq\mu(g)$$

where u is taken as above.

Recall that in (7.40) we have met a specialization $g=f+h \to f$ such that $\mu(f)=\mu(g)$ and $\tau(f)>\tau(g)$. This shows that the equality of Milnor numbers above can actually take place.

(iii) Note that since the closure of a G-orbit is a G-invariant set, we can reformulate Definition (8.7) by asking just $\overline{K^s_{n,p}\cdot j^s f} \ni j^s g$. Then, since the K-orbits are larger than the corresponding R-orbits, it follows that $Rf \to Rg$ implies $Kf \to Kg$. In particular, by (ii) we get $\tau(f)<\tau(g)$. And the inequality for Milnor numbers follows as in (ii) above, using the relation between Milnor numbers and codimensions of R-orbits given by (6.52.i)). In fact, one needs $\mu(f)<\mu(g)$ to deduce that f and g are not K-equivalent and hence to be allowed to write $Kf \to Kg$!

§2. COMPLETE TRANSVERSALS AND SPLITTING LEMMA

To determine the normal forms for the equivalence classes of map germs belonging to some subset $\Sigma \subset E^0_{n,p}$ one usually proceeds inductively. Starting with an (s-1)-jet $u \in J^{s-1}(n,p)$, there are two possibilities:

(i) u *is sufficient*. Then it is easy to decide whether $u \in \Sigma$ and, in the affirmative case, u can be taken as a normal form for its class.

(ii) u *is not sufficient*. Then one first finds normal forms for the s-jets in

$$J^s(u)=(j^{s,s-1})^{-1}(u) \subset J^s(n,p) \quad .$$

If $u_1 \in J^s(u)$ is such a normal form, then one considers the possibilities (i) and (ii) for this s-jet u_1 and so on, until all the orbits in Σ are listed.

Note that in case (ii) one can, more generally, try to find the normal forms for the t-jets in

$$J^t(u)=(j^{t,s-1})^{-1}(u) \subset J^t(n,p), \quad \text{for some} \quad t \geq s.$$

A method for doing this quite efficiently in the case of K-equivalence is by using the *complete transversals* which were introduced in [DG3], [DG4]. We shall present this method here only in its simplest version (i.e. when u is a weighted homogeneous jet with some additional property) which is sufficient for our purposes in this book and is very close in spirit to Proposition (7.41). The general version can be found in [DG4].

Recall from Chapter 6 the definition of the subgroup $G(u) \subset K^t_{n,p}$ and the properties (6.23), (6.24). In analogy to the notion of transversal (recall (4.18) and (4.20)!) we introduce the next definition .

(8.10) DEFINITION

A linear subspace $T^t(u) \subset P^t = \ker j^{t,s-1}$ is called a *complete transversal* if the linear affine space $u+T^t(u)$ intersects all the $G(u)$-orbits in $J^t(u)$.

It will become clear soon that in many concrete cases $\dim T^t(u)$ is much smaller than $\dim J^t(u)$ and hence it is much easier to find normal forms only for the jets in $u+T^t(u)$.

Assume now that the intial jet u is a weighted homogeneous polynomial map of type $(\underline{w},\underline{d})$. Note that we regard u as an $(s-1)$-jet, a t-jet or an element from $H(\underline{w},\underline{d};K) \subset E^o_{n,p}$.

(8.11) PROPOSITION

With the above notations, assume that $\max_j (d_j) < s \cdot \min_i (w_i)$ *where* $j=1,\ldots,p$ *and* $i=1,\ldots,n$.

Then any complement $T^t(u)$ *of the linear subspace* $V^t(u) = T_u(K^t_{n,p} \cdot u) \cap P^t$ *in* P^t *is a complete transversal.*

PROOF

Note that $V^t(u)$ is exactly $T_u(G(u) \cdot u)$ by (6.24). Using now (4.20), it follows that $u+T^t(u)$ intersects transversally all the $G(u)$-orbits in a small neighbourhood U of the jet u in $J^t(u)$.

But using the transformation (7.36) and the assumption on the weights w_i and degrees d_j , it follows that any $G(u)$-orbit in $J^t(u)$ intersects U, and hence $u+T^t(u)$.

In fact, if we take the complement $T^t(u)$ to be generated by "monomials" as in (7.41), it follows that $u+T^t(u)$ is invariant under the transformation (7.36) and hence $u+T^t(u)$ intersects transversally any $G(u)$-orbit in $J^t(u)$ at any point (Exercise). This last property justifies maybe the name of complete transversal.

The first application of (8.11) is the next basic result, which can be regarded as a generalization of Morse Lemma (1.3).

(8.12) PROPOSITION (SPLITTING LEMMA)

A finitely determined function germ $f \in m_n^2$ *is K-equivalent to a polynomial function germ* f_o *of the form*

$$f_o(x_1,\dots,x_n)=Q(x_1,\dots,x_k)\pm x_{k+1}^2\pm\dots\pm x_n^2$$

where k=corank (f), Q *is a polynomial in* k *variables with* ord Q≥3. *Moreover, one has*

$$\mu(f)=\mu(Q) \quad \textit{and} \quad \tau(f)=\tau(Q) \quad .$$

PROOF

Since k=corank (f), it follows that j^2f is a quadratic form of rank n-k (7.44). Now, by the classification of quadratic forms (5.5) one can assume that

$$u=j^2f=\pm x_{k+1}^2\pm\dots\pm x_n^2$$

(Recall that in the complex case there is no need of minus signs!).

We apply now (8.11) with $t\geq O_K(f)$, s=3, p=1, $d_1=2$, $w_i=1$ for i=1,...,n.

Then $V^t(u)=j^t(m_n^2\cdot(x_{k+1},\dots,x_n))$ and hence we can choose $T^t(u)=j^t(m_k^3)$ where $m_k=(x_1,\dots,x_k)$ is regarded as an ideal in $E_k\subset E_n$.

By (8.11) there is a polynomial $Q\in T^t(u)$ such that $u+Q\sim j^tf$. By the choice of t, it follows that f is K-equivalent to f_o=Q+u.

The equality of Milnor and Tjurina numbers follows directly from the Definition (6.49).

We remark finally that there is a similar result to (8.12) for the R-equivalence, see for instance [Gi], p. 125.

§3. CLASSIFICATION OF K-SIMPLE FUNCTION SINGULARITIES

In this section we derive in all the details the K-classification of K-simple complex analytic function germs $f:(\mathbb{C}^n,0)\to(\mathbb{C},0)$.

(8.13) LEMMA

If $k=\text{corank}\ (f)\geq 3$, *then* f *is not* K*-simple.*

PROOF

Let $H=H^3(k,1;\mathbb{C})$, $G=G\ell(k,\mathbb{C})\times G\ell(1,\mathbb{C})$ and $\alpha:G\times H\to H$ be the action defined at (4.12). Recall that (5.11) implies the α-orbits in H coincide to the ρ-orbits. Moreover, for $k\geq 3$ there are infinitely many ρ-orbits in H which meet any nonempty open subset.

Next, using Splitting Lemma (8.12) we can assume that $j^3f=Q+x_{k+1}^2+\ldots+x_n^2$, with $Q\in H$. Then it is easy to see that two 3-jets in $J^3(n,1)$

$$Q+x_{k+1}^2+\ldots+x_n^2 \quad\text{and}\quad Q'+x_{k+1}^2+\ldots+x_n^2\ , \quad\text{with}\quad Q,Q'\in H$$

are $K_{n,1}^3$-equivalent if and only if the cubic forms Q and Q' are in the same α-orbit.

Hence, taking s=3 in Definition (8.1) we see that f is not K-simple.

Therefore, in order to obtain the classification of the K-simple function singularities, we have to check only the germs having corank equal to 0,1 and 2.

(8.14) PROPOSITION

If corank $(f)\leq 1$, *then the function germ* f *is* K*-simple and is* K*-equivalent to the normal form*

$$A_k: f_k=x_1^{k+1}+x_2^2+\ldots+x_n^2$$

where $k=\mu(f)=\mu(f_k)\geq 1$.

PROOF

Let us come back to the main notations from the proof of Splitting Lemma (8.12).

When corank (f)=0, this implies $f\sim f_1$ which is exactly the complex version of Morse Lemma (for K-equivalence, but recall (7.42. iii)).

Assume now that corank (f)=1. Then (8.12) yields

$$f\sim g=Q(x_1)+x_2^2+\dots+x_n^2 \ , \text{ with } Q(x_1)=a_3x_1^3+\dots+a_tx_1^t \quad .$$

Since g is a sufficient jet, it follows that $Q\neq 0$. Let m=ord Q= =min $\{p; a_p\neq 0\}$.

Then clearly $Q(x_1)=x_1^m\cdot h(x_1)$, with $h(0)\neq 0$. This last condition shows that the germ $\bar{h}=h^{1/m}$ is welldefined (up to multiplication with an m-root of the unit).

The coordinate change $\bar{x}_1=x_1\bar{h}(x_1)$, $\bar{x}_i=x_i$ for i>1 gives us $g=\bar{x}_1^m+\bar{x}_2^2+\dots+\bar{x}_n^2$. Hence, if we put m=k+1, we get precisely $f\sim f_k$ for some $k\geq 2$. Moreover $\mu(f)=\mu(f_k)$ by (6.51) and $\mu(f_k)=k$ just by direct computation.

In order to show that this singularity f_k is K-simple and to determine all the singularities which specialize to it, let $U\subset J^s(n,1)$ be a small enough sufficient neighbourhood for f_k $(s\geq k+1)$. To list all the $K^s_{n,1}$-orbits which meet U, we can consider by (4.20) only those orbits which intersect a minimal slice at the point $f_k=j^sf_k$. Since

$$T_{f_k}(K^s_{n,1}\cdot f_k)=j^s(TKf_k)=j^s(m\cdot(x_1^k,x_2,\dots,x_n))$$

we can take as a minimal slice S the set of polynomials

$$f_{a,b}=f_k+a_1x_1+\dots+a_kx^k+b_2x_2+\dots+b_nx_n$$

with $a_i,b_j\in D$=a small disc centered at 0 in $\mathbb{C}$.

If $j^1f_{a,b}=a_1x_1+b_2x_2+\dots+b_nx_n\neq 0$, then clearly $f_{a,b}\sim x_1$, the unique class of nonsingular germs in $m=E^o_{n,1}$.

When $j^1f_{a,b}=0$, then by the first part of the proof we get $f_{a,b}\sim f_j$, where $j=\min\{p, a_p\neq 0\}$ with $p=1,\dots,k,k+1$, $a_{k+1}=1$. This shows that U intersects only the orbits corresponding to the germs $x_1,f_1,\dots,f_k$. By (8.6) it follows that the singularity f_k is $\mathcal{K}$-simple. Moreover, we get the next *diagram of specializations among the* A_k*-singularities*:

(8.15) $\quad A_1 \to A_2 \to \dots \to A_k \to \dots$

(8.16) EXAMPLE

The specialization $A_{k-1} \to A_k$ can be seen very clearly from the family of function germs

$$g_t(x) = tx_1^k + x_1^{k+1} + q \ , \quad \text{where} \quad q = x_2^2 + \dots + x_n^2$$

Then $g_o \in A_k$ and $g_t \in A_{k-1}$ for $t \in \mathbb{C}^*$ (we regard here and in the sequal A_k as being the orbit $\mathcal{K}f_k = \mathcal{R}f_k$ of the normal form f_k in (8.14)). The polynomial *global* function $g_o : \mathbb{C}^n \to \mathbb{C}$ has a single singular point, namely the origin, which is of type A_k.

On the other hand, the *global* function $g_t : \mathbb{C}^n \to \mathbb{C}$ has two singular points, namely the origin of type A_{k-1} and the point $x(t) = (-\frac{tk}{k+1}, 0, \dots, 0)$ of type A_1.

When $t \to 0$, the two singular points 0 and $x(t)$ of the function g_t come together and produce in this way the more complicated singularity of the function g_o.

Note that the two singularities 0 and $x(t)$ sit on distinct level hypersurfaces, i.e.

$$g_t(0) = 0 \neq g_t(x(t)) \ .$$

This behaviour is quite general and the reader should try to work out for himself some other special cases of the specializations discussed in what follows.

Next we discuss the remaining case, corank $(f) = 2$. Such a germ f can be written by Splitting Lemma (8.12) in the form

$$f = Q(x_1, x_2) + q, \quad \text{where} \quad \text{ord } Q \geq 3 \text{ and } q = x_3^2 + \dots + x_n^2 \ .$$

To simplify the notation we put $x = x_1$, $y = x_2$ and forget completely the uninteresting variables $x_3, \dots, x_n$ in most of the considerations.

Then j^3Q is a binary cubic form and by Table (5.8) one can take j^3Q to be one of the normal forms $x^2y + y^3$, x^2y, x^3 or 0.

The first case $j^3Q=x^2y+y^3$ is easy since this jet is K-sufficient (6.18, iii) and hence $f\sim x^2y+y^3$.

Assume now that $j^3Q=x^2y=u$. Then computing $j^t(TK\cdot x^2y)$ it is easy to see that a complete transversal $T^t(u)$ is spanned by $y^4,\ldots,y^t$.

Therefore

$$f\sim x^2y+a_4y^4+\ldots+a_ty^t \text{ , with } t\geq 4 \text{ and } a_i\in\mathbb{C}.$$

Let $k=\min\{p; a_p\neq 0\}$, note that some $a_p\neq 0$ since f is finitely determined and put

$$a_4y^4+\ldots+a_ty^t=y^kh(y).$$

We look for a coordinate change of the special form $\bar{x}=x\cdot a$, $\bar{y}=y\cdot b$ with a,b units in E_2 and such that

$$\bar{x}^2\bar{y}+\bar{y}^k=x^2y+y^kh.$$

A direct computation shows that we can take $b=h^{1/k}$, $a=h^{-1/2k}$ (use $h(0)\neq 0$ to check that everything works well!).

Therefore we have got the equivalence $f\sim x^2y+y^k$, for some $k\geq 4$.

These two cases can be put together in the next statement.

(8.17) PROPOSITION

Let $f=Q(x,y)+q$ *be a finitely determined function germ such that* j^3Q *is a nonzero binary cubic form, different from a perfect cube. Then the germ* f *is* K-*simple and is* K-*equivalent to the normal form*

$$D_k: g_k=x^2y+y^{k-1}+q$$

where $q=x_3^2+\ldots+x_n^2$, $k=\mu(f)=\mu(g_k)\geq 4$.

Before completing the proof of this result we note that the polynomial g_k is weighted homogeneous and hence $Rg_k=Kg_k$ by (7.42). This common orbit is denoted by D_k.

PROOF

The equivalence $f\sim g_k$ was proved before the statement and the equalities $k=\mu(f)=\mu(g_k)$ follow from (6.51) and direct computation (or use (7.27) to compute $\mu(g_k)$!).

To prove that any germ g_k is K-simple we proceed as in the proof of (8.14).

A minimal slice S at the point g_k to the corresponding $K^s_{n,1}$-orbit ($s\geq k$) is formed by the jets of the following type

(8.18) $$g=g_k+\ell(x,y,x_3,\dots,x_n)+q(x,y)+a_3y^3+\dots+a_{k-2}y^{k-2}$$

where ℓ (resp. q) is a linear (resp. quadratic) form and all the coefficients are small enough complex numbers. When $j^1g=\ell\neq 0$, then $g\sim x$, the nonsingular germ. Next, if $j^1g=0$, but $q\neq 0$, it follows that g is of type A_p for some p since corank $(g)\leq 1$ in this case.

And finally, when $\ell=q=0$, then g is clearly of type D_m, for some $m\leq k$ as in the proof above.

In other words, only singularities of type A and D specialize to a D_k-singularity. The next *diagram of specializations among* A *and* D *singularities* makes this more precise.

(8.19)
$$\begin{array}{ccccccccccc} A_1 & \to & A_2 & \to & A_3 & \to & A_4 & \to & A_5 & \to & \dots \\ & & & & & \searrow & & \searrow & & \searrow & \\ & & & & & & D_4 & \to & D_5 & \to & D_6 \to \end{array}$$

In fact the first line is precisely (8.15), the second line is clear by the above proof.

Moreover it is clear that $A_p \not\to D_k$ for $p\geq k$ by (8.9. iii)). Hence it remains just to show that $A_{k-1}\to D_k$ for all $k\geq 4$.

To do this we use the minimal slice S given in (8.18). We have to show that for a suitable choice of the (small!) coefficients in the jet g, we get $g\in A_{k-1}$. Since g_k is a weighted homogeneous polynomial of type (k-2, 2; 2k-2) and since all the other monomials which occur in g have degree strictly less than 2k-2, a transformation of type (7.36) shows that we can make the variable coefficients in g as small as we want without changing the K-orbit. Hence it is enough to determine some coefficients in g such that $g\in A_{k-1}$.

In (8.18) we take $\ell=0$, $q=(x+ty)^2$ for some $t\in\mathbb{C}$ to be later determined. The coordinate change $\bar{x}=x+ty$, $\bar{y}=y$ shows that

$$g \sim x^2(1+y)-2txy^2+t^2y^3+a_3y^3+\ldots+a_{k-1}y^{k-1}$$

with $a_{k-1}=1$.

The new coordinate change

$$\bar{x}=x\sqrt{1+y} - \frac{ty^2}{\sqrt{1+y}}, \ \bar{y}=y \text{ implies}$$

$$g \sim x^2-t^2y^4(1-y+y^2-y^3+\ldots)+t^2y^3+a_3y^3+\ldots+a_{k-1}y^{k-1} .$$

We choose t such that the coefficient of y^{k-1} in this convergent series vanishes. Next we choose $a_3,\ldots,a_{k-2}$ such that the coefficients of $y^3,\ldots,y^{k-2}$ vanish as well. With this choice for the coefficients we get $g \in A_{k-1}$, since the coefficient of y^k above is different from zero.

(8.20) REMARK

A different proof for the specialization $A_{k-1} \to D_k$ can be found in [Lm], p. 212. That proof is surely more quick, but the origin of the good families f_t used there is quite in the dark!

We pass now to the next case, namely when $j^3\Omega$ is a nonzero perfect cube.

(8.21) PROPOSITION

Let $f=Q(x,y)+q$ *be a finitely determined function germ such that* $j^3Q=x^3$. *Then the germ* f *is K-simple if and only if* f *is K-equivalent to one of the normal forms*

$$E_6: f_6=x^3+y^4+q, \quad E_7: f_7=x^3+xy^3+q \quad \text{and} \quad E_8: f_8=x^3+y^5+q .$$

Here $q=x_3^2+\ldots+x_n^2$ *and* $\mu(f_k)=k$ *for* $k=6,7,8$.

Before starting the proof, note again that the polynomials f_k are weighted homogeneous and hence, as in the previous cases, $Rf_k=Kf_k$. And this common orbit is denoted by the symbol E_k, for $k=6,7,8$.

PROOF

A complete transversal for $u=x^3$ is spanned by $y^4,\ldots,y^t$, $xy^3,\ldots,xy^{t-1}$. Hence we can write up to K-equivalence

(8.22) $f=x^3+A(y)+xB(y)$ where $A(y)=a_4y^4+\ldots+a_ty^t$,

$B(y)=b_3y^3+\ldots+b_{t-1}y^{t-1}$.

Let $a=\mathrm{ord}(A)$ and $b=\mathrm{ord}(B)$ be elements in $\mathbb{N}\cup\{\infty\}$ i.e. by convention $a=\infty$ when $A=0$. Clearly one has $a\geq 4$, $b\geq 3$. For simplicity, we forget again about the variables $x_3,\ldots,x_n$.

Then, when $a=4$ we get $j^4f=x^3+a_4y^4+b_3xy^3$ with $a_4\neq 0$. It follows that

$f\in U^+((4,3),12;\mathbb{C})$

and then by (7.39) and (7.45) it follows that $f\sim f_6$, in other words $f\in E_6$.

Assume now $a>4$ and $b=3$. It follows that

$f\in U^+((3,2),9;\mathbb{C})$

and then by (7.39) and (7.45) we get $f\sim f_7$.

When $a=5$ and $b>3$, it follows similarly that $f\sim f_8$.

Finally we show that a function germ f given by (8.22) with $a\geq 6$, $b\geq 4$ is not K-simple. To do this, it is enough to remark that the initial part (with respect to the weights $w_1=2$, $w_1=1$) of such a function f is of the type

$g=x^3+a_6y^6+b_4xy^4$

Now g is an element in $H((2,1),6;\mathbb{C})$ and by (7.32) it follows that the K-orbits of the germs in $U=U((2,1),6;\mathbb{C})$ depend on continuous parameters. Indeed, one has $q_6=1$ in this case and all these orbits have the same Milnor (and Tjurina) number by (7.27). It follows that any neighbourhood of j^sf meets uncountably many of these orbits Kg, for $g\in U$.

We have still to show that these three singularities f_k (k=6,7,8) above are indeed K-simple. This follows from the fact that a slice in a sufficient neighbourhood for the singularity f_k intersects only a finite number of orbits of type A and D by (8.9) as well as the orbits corresponding to the singularities f_p , $6\leq p\leq k$.

The last claim can be made more precise by considering the *diagram of specializations among the* A, D *and* E *singularities*

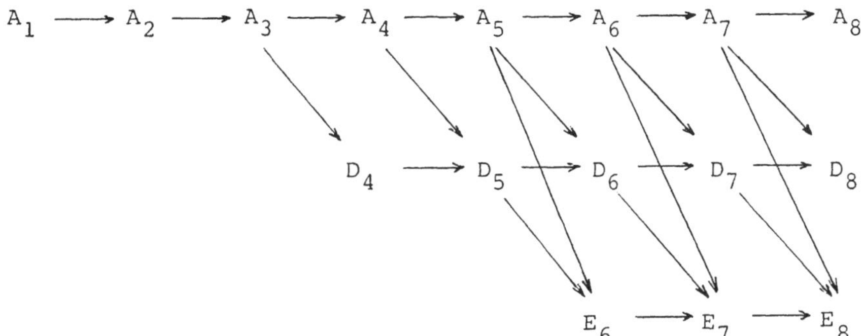

(8.24) DIAGRAM

This diagram is an extension of the diagram (8.19) and the reader can try to derive the additional specializations by the same method as in the proof of (8.19). These computations are more complicated and a real help for him will be to look in [Lm] p. 212-214. See also p. 178 in [Lm] where, as in some other places, the arrows denoting specializations are *reversed by convention*!

To complete the classification of K-simple function singularities, we need the next.

(8.25) LEMMA

A function germ $f=Q(x,y)+q$ *with* ord $Q \geq 4$ *and* $q=x_3^2+\ldots+x_n^2$ *cannot be K-simple.*

PROOF

Similar to (8.13) since the α-action on $H^4(2,1;\mathbb{C})$ has infinitely many orbits. Roughly speaking, a general form $u \in H^4(2,1;\mathbb{C})$ determines four points in $\mathbb{P}^1$ by the equation $u=0$ and the α-orbit of u is determined by the cross-ratio of these points, a continuous parameter.

We sum up our classification in the next statement.

(8.26) THEOREM

A K-simple function singularity $f:(\mathbb{C}^n,0) \to (\mathbb{C},0)$ *is K-equivalent to one of the next normal forms*

A_k : $x^{k+1}+q$, $k\geq 1$

D_k : $x^2y+y^{k-1}+q$, $k\geq 4$

E_6 : x^3+y^4+q, E_7 : x^3+xy^3+q, E_8 : x^3+y^5+q

Here q *denotes the sum of the squares of the other variables and the inferior indices give the corresponding Milnor numbers.*

The reader should compare this list of normal forms with Corollary (7.45) and remark the use of the same symbols for the same singularities. Thus one may say that the property $D(\underline{w},d)<d$ from (7.45) characterizes the K-simple singularities in the class of weighted homogeneous singularities.

§4. CLASSIFICATION OF R-SIMPLE FUNCTION SINGULARITIES

In this section we show that the R-simple function singularities $f:(\mathbb{C}^n,0) \to (\mathbb{C},0)$ coincide to the K-simple ones listed in Theorem (8.26).

(8.27) THEOREM (ARNOLD)

A function singularity $f:(\mathbb{C}^n,0) \to (\mathbb{C},0)$ *is R-simple if and only if the singularity* f *is K-simple. In the affirmative case,* f *is R-equivalent to one of the normal forms* $A_k (k\geq 1)$, D_k $(k\geq 4)$, E_6 , E_7 *and* E_8 *listed in* (8.26).

FIRST PROOF

It follows from Definition (8.1) that if f is R-simple, then f is K-simple. Moreover, since all the K-simple function singularities are weighted homogeneous by (8.27), it follows that their K-orbits coincide to the corresponding R-orbits. Hence any K-simple singularity has a sufficient neighbourhood which intersects finitely many orbits (all corresponding to some K-simple singularities) and therefore it is also R-simple.

SECOND PROOF (without using (8.27)).

Using the foliation of a sufficient K-orbit by the corresponding $\mathcal{R}$-orbits (6.55), it follows that an $\mathcal{R}$-simple function singularity f satisfies $\mu(f)=\tau(f)$. Hence f is $\mathcal{R}$-equivalent to a weighted homogeneous polynomial f_o by Saito's result (7.42). Moreover f_o can be taken in the normalized form from (7.43).

Assume that the homogeneity type $(\underline{w},d)$ of f_o satisfies the inequality $D(\underline{w},d)\geq d$. Then as in the proof of (7.39) we get a family $f_t=f+tu$ and for t near 0 this family contains infinitely many $\mathcal{R}$-equivalence classes. Indeed $\mu(f_t)=\mu(f_o)=$constant implies that these $\mathcal{R}$-orbits have the same codimension. And they do not coincide since $u\notin T\mathcal{R}f_o$ by its choice. Moreover the family f_t does not define a specialization to f_o by (8.9 iii)).

Since f_o is $\mathcal{R}$-simple, it follows that the above case cannot occur. Then one has $D(\underline{w},d)<d$ and hence f_o is $\mathcal{R}$-equivalent to one of the normal forms in (7.45).

And a case-by-case analysis using slices as in the proof of (8.27) shows that all these normal forms define $\mathcal{R}$-simple function singularities. Compare with [AGV], p. 218.

(8.28) REMARKS

(i) With a little additional effort to analyse the signs during the reduction to normal forms, one can similarly obtain the *$\mathcal{R}$-classification of $\mathcal{R}$-simple smooth function germs* $f:(\mathbb{R}^n,0)\to(\mathbb{R},0)$. The corresponding normal forms are the following:

$$A_{2k}:x^{2k+1}+q;\ A^{\pm}_{2k-1}:\pm x^{2k}+q\quad(k\geq 1);\ D^{\pm}_k:x^2y\pm y^{k-1}+q\quad(k\geq 4)$$

$$E^{\pm}_6:x^3\pm y^4+q;\ E_7:x^3+xy^3+q;\ E_8:x^3+y^5+q$$

where q is a sum (with $\pm$ signs) of the squares of the other variables.

(ii) In Chapter 10 we shall give some reasons for the use of the symbols A, D, E to denote simple function singularities. More precisely, we shall relate the minimal resolutions of surface singularities of type A, D, E to the corresponding Dynkin diagrams of irreducible root systems, a classical topic in Lie Algebras theory, see for instance [H1] p. 58.

(iii) Assume for simplicity that we are in the complex case and n=3 (variables x,y and z). Then the singularities which are the next ones in complexity after the simple singularities are the following one-parameter families:

$\widetilde{E}_6$: $x^3+y^3+z^3+3txyz$, $t \in \mathbb{C}$, $t^3 \neq -1$

$\widetilde{E}_7$: $x^4+2tx^2y^2+y^4+z^2$, $t \in \mathbb{C}$, $t \neq \pm 1$

$\widetilde{E}_8$: $x^3+3txy^4+y^6+z^2$, $t \in \mathbb{C}$, $4t^3 \neq -1$

These singularities form the *border line* between the simple singularities A-D-E and the other function singularities: any finitely determined function singularity $f:(\mathbb{C}^3,0) \to (\mathbb{C},0)$ which is not simple has (arbitrarily small) sufficient neighbourhoods which intersects infinitely many R (or K) orbits corresponding to germs in at least one of the families $\widetilde{E}_6$, $\widetilde{E}_7$ or $\widetilde{E}_8$.

To see this, just note that (essentially) the family $\widetilde{E}_6$ has occured in (8.13), the family $\widetilde{E}_7$ in (8.25) and $\widetilde{E}_8$ in the proof of (8.21) as function q.

A singularity of type $\widetilde{E}_6$, $\widetilde{E}_7$ or $\widetilde{E}_8$ is called a hypersurface *simple-elliptic singularity* and this class of singularities was carefully studied by K. Saito [S2].

We just note here that they are all weighted homogeneous singularities and are characterized in the class of weighted homogeneous function singularities by the condition $D(\underline{w},d)=d$, where $(\underline{w},d)$ is the homogeneity type.

The corresponding Milnor numbers are

$\mu(\widetilde{E}_k)=k+2$ for $k=6,7,8$.

(8.29) COROLLARY

Any function singularity f *with* $\mu(f) \leq 7$ *is* R *and* K *simple.*

CHAPTER 9

CLASSIFICATION OF SIMPLE 0-DIMENSIONAL COMPLETE INTERSECTIONS

§1. FURTHER INVARIANTS FOR SINGULARITIES: BOARDMAN SYMBOL AND HILBERT-SAMUEL FUNCTION

In this Chapter we present the K-classification of K-simple equidimensional map germs $f:(\mathbb{C}^n,0) \to (\mathbb{C}^n,0)$. Such a map germ f is finitely K-determined exactly when the fiber $f^{-1}(0)$ is 0-dimensional, and hence a complete intersection . This remark explains the title of this Chapter.

The listing of the normal forms in this situation is due to Giusti [Gt], but the complete list of their specializations was later found by the author and Gibson [DG1], [DG2] (compare to [AGV], p. 172 and [Ln] p.130).

We start this section by introducing an important numerical invariant or, rather, a sequence of numerical invariants for any map germ $f:(K^n,0) \to (K^p,0)$, namely the Boardman symbol. For any finitely generated ideal $I \subset E_n$ we define the s-th *Jacobian extension* of I to be the ideal $\Delta^s I = I + I'$, where I' is the ideal in E_n generated by all $(n-s+1)\times(n-s+1)$ minors of the Jacobian matrix

$$\left(\frac{\partial f_i}{\partial x_j}\right) \begin{array}{l} i=1,\dots,p \\ j=1,\dots,n. \end{array}$$

with $f_1,\dots,f_p$ being a system of generators for the ideal I.

Elementary Linear Algebra shows that one has a sequence of inclusions

$$(9.1) \qquad I=\Delta^o I \subset \Delta^1 I \subset \dots \subset \Delta^n I$$

Suppose now that the ideal I is *proper* (i.e. $I \neq E_n$).

Then the *critical Jacobian extension* of I is the last ideal $\Delta^{i_1} I$ in the sequence (9.1) which is proper. This ideal

$\Delta^{i_1}I$ has in turn its critical Jacobian extension $\Delta^{i_2}\Delta^{i_1}I$ and so on. The sequence of nonnegative integers $(i_1,i_2,\dots)$ obtained in this way is called the *Boardman symbol of the ideal* I.

(9.2) DEFINITION

The *Boardman symbol of a map germ* $f\in E^{o}_{n,p}$ is the Boardman symbol of the ideal I_f generated by the components $f_1,\dots,f_p$ of f in E_n.

Sometimes the Boardman symbol $(i_1,i_2,\dots)$ of the map germ f is denoted by $\Sigma f=\Sigma^{i_1,i_2,\dots}$ and the germ f is called a *map germ of type* $\Sigma^{i_1,i_2,\dots}$.

If we take only the first k integers in the Boardman symbol of a map germ we obtain the kth *order Boardman symbol* $\Sigma^{i_1,\dots,i_k}$.

(9.3) EXERCISE

Show that for the simple function singularities in (8.26) one has:

$\Sigma(A_k)=(n,1,\dots,1,0,0,\dots)$ with (k-1)-repeated 1's.

$\Sigma(D_k)=(n,2,0,0,\dots)$

$\Sigma(E_6)=\Sigma(E_7)=(n,2,1,0,0,\dots)$

$\Sigma(E_8)=(n,2,1,1,0,0,\dots)$.

Hint: direct computations using the normal forms in n variables given in (8.26).

Note that all the D_k singularities have the same Boardman symbol. This indicates that this symbol taken alone is a poor invariant, i.e. it is unable to distinguish among many nonequivalent germs.

On the other hand, consider the next Table (9.4) which contains the Milnor numbers and the third order Boardman symbols of the simple function singularities.

(9.4) TABLE

Type	μ	Σ^{i_1,i_2,i_3}
A_1	1	(n,0,0)
A_2	2	(n,1,0)
A_k	k	(n,1,1) , $k\geq 3$
D_k	k	(n,2,0)
E_6	6	(n,2,1)
E_7	7	(n,2,1)
E_8	8	(n,2,1)

A look convinces the reader that any two distinct simple function singularities have distinct values (of at least one) of the invariants considered in this Table. A more subtle observation and a challenging exercise for the reader is the following.

(9.5) EXERCISE

Consider a function germ $f:(\mathbb{C}^n,0) \to (\mathbb{C},0)$ which has the same Milnor number and third order Boardman symbol as a simple germ (say f_o) in Table (9.4). Then f is a simple singularity of the same type as f_o.

For more details on the Boardman symbol of a map germ and for a proof of the next basic result, we send the reader to [Gi].

(9.6) PROPOSITION

(i) *The Boardman symbol of a map germ is a contact invariant, i.e. two K-equivalent map germs have the same Boardman symbol.*

(ii) *The* k^{th} *order Boardman symbol of a map germ* f *depends only on the* k-*jet* $j^k f$.

There is also a geometric approach to the Boardman symbol under some genericity conditions (for details we refer to [GG]), based essentially on the next easy result.

(9.7) EXERCISE

The first order Boardman symbol of a map germ f is Σ^k, where $k=\dim \ker df(0)$.

We pass now to the second invariant we like to introduce in this section: the Hilbert-Samuel function (compare to [Ha], p. 394).

Let I be an ideal in E_n and let $A=E_n/I$ be the quotient local ring with maximal ideal $m_A=m/I$.

(9.8) DEFINITION

The *Hilbert-Samuel function of the local ring* A (or of the corresponding *ideal* I) is the function

$$H_A(k)=H_I(k)=\dim_K\left(\frac{A}{m_A^k}\right)=\dim_K\left(\frac{E_n}{I+m^k}\right)$$

defined for any integer $k>0$.

This function is related to a Poincaré series as in (7.20) in the following way. Consider the graded ring associated to A with respect to the maximal ideal m_A, namely

$$(9.9) \qquad Gr(A)=\bigoplus_{i\geq 0}\frac{m_A^i}{m_A^{i+1}}=\bigoplus_{i\geq 0}\frac{I+m^i}{I+m^{i+1}}$$

Let

$$(9.10) \qquad h_A(k)=h_I(k)=\dim_K Gr(A)_k=\dim_K \frac{I+m^k}{I+m^{k+1}}$$ be the coefficients of the corresponding Poincaré series. Then it is clear that

$$(9.11) \qquad H_I(k)=h_I(0)+\dots+h_I(k) \quad \text{for all } k>0.$$

When $I=I_f$ is the ideal corresponding to a map germ $f\in E^o_{n,p}$, then we denote H_I, h_I simply by H_f, h_f and call H_f the

Hilbert-Samuel function of the map germ f.

(9.12) EXERCISE

(i) Show that H_f is a contact invariant of the map germ f in the sense of (9.6.i) .

(ii) Show that for any ideal $I \subset E_n$ one has

$$\frac{I+m^k}{I+m^{k+1}} = \frac{m^k}{I \cap m^k + m^{k+1}}$$

Our interest in the Hilbert-Samuel function is based on the following connection with specializations.

(9.13) PROPOSITION

Let $f,g \in E^o_{n,p}$ *be map germs such that* $Kf \to Kg$. *Then* $H_f(k) \leq H_g(k)$ *for any* $k \geq 1$.

PROOF

Note the obvious equality

$$H_f(k) = \dim \frac{E_n}{m_n^k} - \dim \frac{I_f + m^k}{m^k}$$

The first vector space dimension is independent of f, while the second is the rank of the matrix having as rows the coefficients of the polynomials $x_1^{a_1} \dots x_n^{a_n} \cdot f_i$ modulo m^k, with $i=1,\dots,p$ and $a_1+\dots+a_n<k$. Then the above statement follows from the well known semicontinuity property of the rank of a matrix.

(9.14) REMARK

The corresponding statement for the coefficients $h_f(k)$, $h_g(k)$ of the Poincaré series does not hold in general as shown by Exercise (9.46) below.

In the next section we shall be mainly concerned with

ideals $I \subset m^2 \subset E_2$ and for them the coefficients $h_I(k)$ satisfy the following additional property, due to Iarrobino [I].

(9.15) LEMMA

For an ideal $I \subset m^2 \subset E_2$ *one has*

$$2=h_I(1) \geq h_I(2) \geq \ldots \geq h_I(k) \geq \ldots$$

PROOF

Let R_j be the vector space of homogeneous polynomials in x and y of degree j. Using (9.12.ii) it follows that it is enough to prove the following. For any vector subspace $V \subset R_j$, $V \neq 0$ one has $\dim (R_1 \cdot V) \geq \dim V + 1$, where $R_1 \cdot V$ is the vector subspace generated in R_{j+1} by all the products rv with $r \in R_1$ and $v \in V$.

To show this, choose a basis $f_1, \ldots, f_p$ of V such that f_1 has a monomial term higher (or equal) in y-degree than any term in $f_2, \ldots, f_p$. Then yf_1, $xf_1, \ldots, xf_p$ are linearly independent in R_1V and this ends the proof.

(9.16) NOTATION

When $I=I_f$ for a finitely determined map germ $f:(K^2,0) \to (K^2,0)$ we denote by $d(f)$ (resp. $e(f)$) the number of times the number 2 (resp. 1) occurs in the decreasing sequence from (9.15).

It is clear that these two integer invariants contain exactly the same information as the whole Hilbert-Samuel function H_f.

§2. NORMAL FORMS AND SPECIALIZATIONS

In the K-classification of map germs, the next easy result (which can be regarded as an analogue of the Splitting Lemma) is quite useful.

(9.17) LEMMA

A map germ $f \in E^o_{n,p}$ *of type* Σ^k *is* K*-equivalent to a map germ* $g \in E^o_{n,p}$ *of the form*

$$g(x_1,\ldots,x_n)=(x_1,\ldots,x_{n-k},g_{n-k+1}(\bar{x}),\ldots,g_p(\bar{x}))$$

where $\bar{x}=(x_{n-k+1},\ldots,x_n)$ *and* $j^1g_i=0$ *for all* $i=n-k+1,\ldots,p$.

PROOF

Recall the proof of (3.13) or, alternatively, use the method of complete transversals as in the proof of the Splitting Lemma (assuming this time f finitely K-determined!)

As a consequence, it is enough to classify only map germs f whose components f_i are all in the square m^2 of the maximal ideal m.

From now on we restrict our attention to such germs which in addition are complex analytic and equidimensional (i.e. n=p).

Our aim is thus to obtain the K-classification of the K-simple germs in the set

$$S_n=\{f:(\mathbb{C}^n,0)\to(\mathbb{C}^n,0);\ \Sigma f=\Sigma^n\}$$

(9.18) LEMMA

For $n\geq 3$, *there is no* K*-simple germ in* S_n.

PROOF

At the 2-jet level, one has to consider the α-action from (4.12) with $G=G\ell(n,\mathbb{C})\times G\ell(n,\mathbb{C})$ and $H=H^2(n,n;\mathbb{C})$. Then we obviously have

$$\dim G=2n^2 \quad \text{and} \quad \dim H=\tfrac{1}{2}n^2(n+1)\ .$$

Moreover, the 1-parameter subgroup in G given by $(t\cdot I,t^2\cdot I)$ where $I\in G\ell(n,\mathbb{C})$ denotes the unit matrix, acts trivially on H.

Hence, for any element $u\in H$ one has

$$\dim(G\cdot u)\leq \dim G-1<\dim H \quad \text{for} \quad n\geq 3.$$

This shows that any nonempty open subset in H intersects infinitely many G-orbits.

Since $H=J^2(n,n)$ and the G-orbits coincide to the $K^2_{n,n}$-orbits, the result follows.

(9.19) REMARK

When n=3 one can obtain a 1-parameter family of G-orbits in the following way.

For a function $f \in m_n^2$, we define its gradient

$$\text{(9.20)} \qquad \text{Grad}\,(f):(K^n,0) \to (K^n,0), \quad x \to \left(\frac{\partial f}{\partial x_1}(x),\ldots,\frac{\partial f}{\partial x_n}(x)\right) .$$

Let $f_t=x^3+y^3+z^3+3txyz$ be the normal form for the family $\tilde{E}_6$. Then we leave as an exercise to the reader to show that the family Grad (f_t), for $t \in \mathbb{C}$ meets infinitely many G-orbits in H, in the notations from (9.18).

Therefore, by (9.18) we have only to classify the simple singularities in S_1 and S_2. The first case is trivial.

(9.21) LEMMA

A finitely K-determined map germ $f \in S_1$ is K-simple and is K-equivalent to the normal form $A^k:x^{k+1}$ where $k=\text{codim}\,(f)\geq 1$.

It is usual to identify a map germ $f \in E^o_{n,p}$ with its *q-suspension* $\bar{f} \in E^o_{n+q,p+q}$ given by

$$\bar{f}(x,y)=(f(x),y), \quad \text{where} \quad y=(y_1,\ldots,y_q) .$$

In fact, suspension is the inverse of the reduction described in (9.17).

(9.22) EXERCISE

Show that $\Sigma\bar{f}=\Sigma f$ and $\text{codim}\,(\bar{f})=\text{codim}\,(f)$, for any $q\geq 1$.

In particular, we also write A^k for the map germ $(x,y) \to (x^{k+1},y)$ in $E^o_{2,2}$. Among these singularities A^k, we have the next obvious specializations

$$\text{(9.23)} \qquad A^1 \to A^2 \to A^3 \to \ldots .$$

To start the classification of the map germs in S_2 , we make the list of their 2-jets and corresponding second order Boardman symbols (use (5.21) and (9.6.ii) :

(9.24) TABLE

2-jet	Σ^{i_1,i_2}
(x^2,y^2)	(2,0)
(x^2,xy)	(2,0)
$(xy,0)$	(2,0)
$(x^2,0)$	(2,1)
$(0,0)$	(2,2)

Hence there are three distinct cases to be treated. The case of $\Sigma^{2,0}$-germs is quite easy.

(9.25) PROPOSITION

A finitely K-determined map germ $f \in S_2$ *which is of type* $\Sigma^{2,0}$ *is K-simple and is K-equivalent to the normal form*

$B^{p,q}: (xy, x^p+y^q)$ *with* $q \geq p \geq 2$.

PROOF

Note that we denote a map germ $f \in E^o_{2,2}$ by (f_1, f_2), indicating thus the components.

We present the proof only in the most complicated case, namely when $j^2f=(xy,0)$. The other cases are left to the reader.

When $j^2f=(xy,0)$, a complete transversal is spanned by $(0,x^3),\dots,(0,x^t)$, $(0,y^3),\dots,(0,y^t)$. Hence we can write up to K-equivalence

$f=(xy, x^pA(x)+y^qB(y))$

where the polynomials A,B satisfy $A(0)\cdot B(0)\neq 0$. Indeed, the finite determinacy of f shows that one cannot have A=0 or B=0.

A coordinate change of the type $\bar{x}=x\cdot a(x)$, $\bar{y}=y\cdot b(y)$ with a,b units shows that f is equivalent to the normal form $B^{p,q}$ (compare to (8.17)).

A first natural question is the next: *Is it true that* $B^{p,q}=B^{s,t}$ *if and only if* p=s, q=t? In other words: *do the above normal forms really correspond to different K-orbits?*

To answer such a question we compute the numerical invariants introduced in this book. By simple direct computations one gets

(9.26) $\quad \Sigma(B^{p,q})=(2,0,0,\ldots)$

$$\text{codim }(B^{p,q})=p+q,\ d(B^{p,q})=p-1,\ e(B^{p,q})=q-p+1$$

For two distinct pairs $(p,q)\neq(s,t)$, with $q\geq p\geq 2$, $t\geq s\geq 2$ note that either $d(B^{p,q})\neq d(B^{s,t})$ or $e(B^{p,q})\neq e(B^{s,t})$. Hence $B^{p,q}\neq B^{s,t}$ and this answers our question above.

To end the proof of (9.25) it remains to show that any $B^{p,q}$ singularity is K-simple. This follows essentially from the next more precise result.

(9.27) PROPOSITION

The singularities in $E^{o}_{2,2}$ *which specialize to* $B^{p,q}$ *are the following*

(i) $B^{s,t}$ *for* $s\leq p$ *and* $t\leq q$.

(ii) A^{k} *for* $k<p+q$.

Before giving the proof, we note that from this point on we leave the reader to draw the other specialization diagrams, which will be increasingly complicated (if he enjoys this job!).

PROOF

The point (i) is clear since a slice at $f=(xy,x^{p}+y^{q})$ in the subspace of jets g with $j^{1}g=0$ (corresponding to the subset S_2) is given by

$$g=(g_1,g_2)=(xy,a_2x^2+\ldots+a_px^p+b_2y^2+\ldots+b_qy^q)$$

where $a_i,b_j\in\mathbb{C}$, $a_p=b_q=1$.

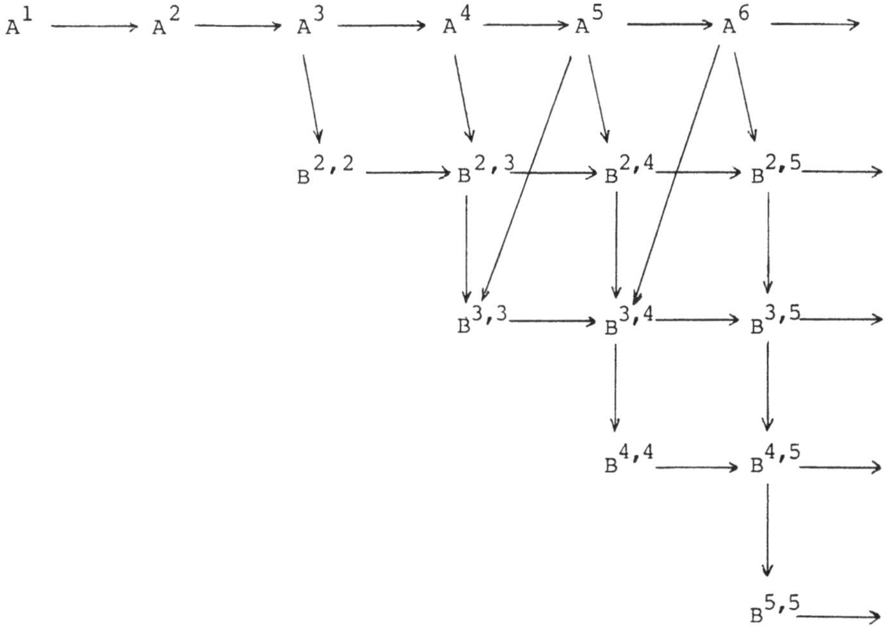

SPECIALIZATION DIAGRAM

(ii) The condition $k<p+q$ is a necessary condition for $A^k \to B^{p,q}$ by (8.9.i) since $k=\text{codim}\,(A^k)$ and $p+q=\text{codim}\,(B^{p,q})$.

To prove the sufficiency, it is enough to show

$$A^{p+q-1} \to B^{p,q} \tag{9.28}$$

A slice S at f in the whole jet space $J^t(2,2)$, for large enough t, is formed by the elements

$$\bar{g}=(ux+vy+g_1\ ,\ a_1x+b_1y+g_2)$$

with $u,v\in\mathbb{C}$ and g_1,g_2 as above.

Let us take $u=1$ and $\bar{x}=x-wy+xy$, with $w=-v$. Then clearly $x=wy(1+y)^{-1}\text{mod}\ \bar{x}$. It is enough hence to show that

(9.29) $$\sum_{i=1,p} a_i w^i y^i (1+y)^{-i} + \sum_{j=1,q} b_j y^j = cy^{p+q} \mod y^{p+q+1}$$

for some constants $a_i, b_i, w, c \in \mathbb{C}$, $a_p = b_q = 1$, $c \neq 0$.

By multiplying (9.29) with $(1+y)^p$ we get the next equivalent problem (which is simpler, being in the polynomial ring $\mathbb{C}[y]$!):

(9.30) $$\sum_{i=1,p} a_i w^i y^i (1+y)^{p-i} + \sum_{j=1,q} b_j y^j (1+y)^p = cy^{p+q}$$

Note that the polynomials $y^i(1+y)^{p-i}$ and $y^j(1+y)^p$ which occur above form a basis of the vector space

$$\{P \in \mathbb{C}[y];\ P(0)=0,\quad \deg P \leq p+q\}$$

It follows that

$$\sum_{i=1,p} \bar{a}_i y^i (1+y)^{p-i} + \sum_{j=1,q} \bar{b}_j y^j (1+y)^p = y^{p+q}$$

for some coefficients $\bar{a}_i, \bar{b}_j$.

Moreover $\bar{b}_q = 1$ and $\bar{a}_p \neq 0$ by obvious reasons. Then the data: $w = (\bar{a}_p)^{1/p}$, $a_i = \bar{a}_i \cdot w^{-i}$, $b_j = \bar{b}_j$ and $c=1$ satisfy (9.30) and hence the proof is ended. Note that we have used implicitly the fact that f is weighted homogeneous of type $\underline{w} = (q,p)$, $\underline{d} = (p+q, pq)$ and that all the additional monomials in g have smaller degrees. This allows us to use any coefficients in (9.30) and not just sufficiently small ones (compare to the proof of (8.19)).

This remark will be in order for all the other specializations proofs below.

Now we consider the third case in (9.24).

(9.31) LEMMA

There is no K-simple map germ of type $\Sigma^{2,2}$ *in* S_2.

PROOF

For a $\Sigma^{2,2}$-germ f, its 3-jet $j^3 f$ can be regarded as a pencil of binary cubic forms. Then by (5.22) it follows that the classification of the orbits around $j^3 f$ cannot be discrete.

(9.32) REMARK

Let $f_t=x^4+2tx^2y^2+y^4$, $t\in\mathbb{C}\setminus\{\pm 1\}$ be the normal form for the family $\widetilde{E}_7$ in two variables (i.e. we forget the innocuous square z^2 which appears in (8.28 iii)). Then the map germ Grad (f_t) produces a 1-parameter family of pencils of binary cubics, different from the family which appears in (5.22).

From now on we concentrate on the remaining case of $\Sigma^{2,1}$-germs in S_2 which is more subtle. For such a germ f, we can assume that $j^2f=(x^2,0)$. A complete transversal in $J^t(2,2)$ is spanned by $(y^i,0)$, $(0,xy^{i-1})$ and $(0,y^i)$ where $i=3,\dots,t$.

Hence our problem reduces to find the K-simple germs among the map germs $f\in S_2$ of the form

$$f=(x^2+y^kA(y),xy^\ell B(y)+y^mC(y))$$

where the polynomials A,B,C satisfy $A(0)\cdot B(0)\cdot C(0)\neq 0$, $k,m\geq 3$ and $\ell\geq 2$ (we leave as an exercise the case when A, B or C is 0!).

The coordinate change $\overline{x}=xa(y)$, $\overline{y}=y$ with $a^2=A$ yields the simpler *prenormal form*

$$(9.33)\qquad f=(x^2+y^k,\ xy^\ell+y^mR(y))$$

where $R\in E_1$ with $R(0)\neq 0$, $k\geq 3$, $m\geq 3$, $\ell\geq 2$.

This prenormal form can be further simplified.

(9.34) LEMMA

(i) *If* $m\leq\ell$ *then* f *is* K-*equivalent to the normal form*

$$C^m:(x^2,y^m),\ m\geq 3 \text{ if } k\geq m$$

and to the normal form

$$E^{k,m}:(x^2+y^k,y^m)\quad m>k\geq 3\quad \text{if}\quad k<m\ .$$

(ii) *If* $m\geq 2\ell$, *then* f *is* K-*equivalent to the normal form*

$$F^{p,\ell}:(x^2+y^p,xy^\ell)\quad p\geq 3,\quad \ell\geq 2$$

with $p=\mathrm{ord}(y^k+R^2\cdot y^{2(m-\ell)})$.

PROOF

Assertion (i) is obvious and for (ii) make the coordinate change $\bar{x}=x+R\cdot y^{m-\ell}$.

The situation we have now reached is exactly the same as when we have classified the simple function germs g in E_2 with $j^3g=x^3$ in Proposition (8.21). Namely, we are going to produce the simplest 1-parameter family of singularities of the form (9.33) and use it to derive strong restrictions on the exponents k,ℓ,m in order that the map germ f be K-simple.

To find this family which forms the border line between the simple and the non-simple $\Sigma^{2,1}$-map germs in S_2, we start with the remaining simple-elliptic singularity $\tilde{E}_8$.

Let $g_t=x^3+3txy^4+y^6$, $t\in\mathbb{C}$, $t^3\neq-\frac{1}{4}$ be the corresponding normal form after forgetting again the innocuous square z^2. Then the map germs

$$G_t=\frac{1}{3}\mathrm{Grad}\ (g_t)=(x^2+ty^4,\ 4txy^3+2y^5)$$

define a 1-parameter family of (distinct) K-orbits. Indeed a direct computation shows that

$$\frac{d}{dt}G_t=(y^4,4xy^3)\notin TKG_t \quad \text{for any} \quad t\in\mathbb{C}.$$

To simplify this computation, use the weighted homogeneity with respect to $\underline{w}=(2,1)$.

This 1-parameter family G_t and an argument similar to that in (8.21) - this time in the vector space $H((2,1),(4,5);\mathbb{C})$ - implies that the map germ f in (9.33) cannot be simple if $k\geq4$, $\ell\geq3$ and $m\geq5$.

The remaining cases produce indeed simple singularities and their classification is contained in the next result.

(9.35) PROPOSITION

A $\Sigma^{2,1}$-map germ f in S_2 is K-simple if and only if f is K-equivalent to one of the following normal forms

Symbol	Normal form	codim	d	e
$C^k;k=3,4$	(x^2,y^k)	3k-2	k-1	1
$E^{3,k};k>3$	(x^2+y^3,y^k)	2k+1	k-1	1
$F^{3,k};k>1$	(x^2+y^3,xy^k)	2k+4	k	2
$F^{k,2};k>3$	(x^2+y^k,xy^2)	k+5	2	k-1

We note, before giving the proof, that the invariants codim, d and e in this table show that these K-orbits are indeed *distinct*.

PROOF

By the previous discussion, we have only to analyse what happens when at least one of the inequalities $k\geq 4$, $\ell\geq 3$ and $m\geq 5$ does not hold for the map germ f in (9.33).

When k=3 one obtains easily the normal forms C^3, $E^{3,m}$ and $F^{3,m}$ by (9.34).

When $k\geq 4$ and $\ell=2$ one gets the $F^{m,2}$ series of singularities, $F^{3,2}$ and C^3.

When $k\geq 4$, $\ell\geq 3$ and m=3,4 one gets only the orbits C^3 and C^4.

And the classification process itself shows that all these singularities are K-simple.

In conclusion, the K-simple singularities in $E^o_{2,2}$ (or, by suspension, in any $E^o_{n,n}$ with $n\geq 2$) consist of five infinite series A^*, $B^{*,*}$, $E^{3,*}$, $F^{3,*}$ and $F^{*,2}$ and two exceptional cases C^3 and C^4. In fact we can write $C^3=E^{3,3}$ and integrate in this way the singularity C^3 into the series $E^{3,*}$. But no such thing is possible for the remaining singularity C^4.

In analogy with Exercise (9.5), we state the following.

(9.36) COROLLARY

Let $f, f_o \in E^o_{2,2}$ *be two map germs such that* f_o *is K-simple. Assume that the second order Boardman symbols of* f *and* f_o *coincide and that* $\operatorname{codim}(f) = \operatorname{codim}(f_o)$, $d(f) = d(f_o)$ *and* $e(f) = e(f_o)$.

Then f *is K-equivalent to the simple germ* f_o.

PROOF

It is enough to look carefully into the classification of K-simple germs. Alternatively, one can consult [DG3] where all the $\Sigma^{2,1}$-germs in $E^o_{2,2}$ are classified up to K-equivalence.

We end this section by describing the singularities which specialize to each $\Sigma^{2,1}$ simple singularity.

(9.37) PROPOSITION

The singularities in $E^o_{2,2}$ *which specialize to the singularity* $E^{3,k}$ *are the following*

A^m *for* $m<2k$; $B^{p,q}$ *for* $p+q \leq 2k$; C^3;

$F^{3,m}$ *for* $2 \leq m < k-1$; $E^{3,m}$ *for* $4 \leq m < k$.

PROOF

First we treat the case of $B^{p,q}$ singularities. Since $\operatorname{codim}(B^{p,q}) = p+q$ and $\operatorname{codim}(E^{3,k}) = 2k+1$, the condition $p+q<2k+1$ is necessary by (8.9).

To prove the sufficiency, note the following *geometric description* of the singularity $B^{p,q}$.

Let $f=(f_1,f_2)$ be a map germ in $E^o_{2,2}$ such that the curve singularity $f_1=0$ consists of two smooth branches, meeting transversally. Then $f \in B^{p,q}$ exactly when the restrictions of the second component f_2 to these branches vanish at the origin $0 \in \mathbb{C}^2$ with multiplicities p and q respectively.

A slice to $E^{3,k}$ in the subspace of jets u with $j^1u=0$ is given by the formula

$$(9.38) \qquad f = (x^2+y^3+ay^2, \sum_{i=1,k-2} a_i x y^i + \sum_{j=2,k} b_j y^j)$$

with $a, a_i, b_j \in \mathbb{C}$, $b_k = 1$.

Take a=-1 and note that the branches at 0 of the curve $x^2+y^3-y^2=$ $=0$ are parametrized by $x=t(1-t^2)$, $y=1-t^2$ with t in a neighbourhood of the points 1 and -1 respectively (recall Figure (1.4)).

By the remark above and by (9.27.i), we have only to show that for any $p,q\geq 2$, $p+q=2k$ one can find $a_i, b_j \in \mathbb{C}$ such that

$$\sum_{i=1,k-2} a_i t(1-t^2)^{i+1} + \sum_{j=2,k} b_j (1-t^2)^j = (-1)^k (t+1)^p (t-1)^q$$

If we divide out the common factor $(1-t^2)^2$, we get $(2k-3)$ linearly independent polynomials

$$t(1-t^2)^i \text{ for } i=0,\dots,k-3; \quad (1-t^2)^j \text{ for } j=0,\dots,k-2$$

in the vector space

$$V=\{P\in\mathbb{C}[t];\ \deg P\leq 2k-4\} .$$

Since $\dim V=2k-3$, these polynomials form a basis of V and hence one can find unique a_i, b_j satisfying the equality above. Moreover it is clear that $b_k=1$ and hence the proof in this case is complete.

For the A^m singularities the condition $m<2k$ is sufficient by (9.27.ii) and the above result.

For necessity we use the Hilbert-Samuel functions as in (9.13). Indeed, one has

$$m+1=H_{A^m}(s)\leq H_{E^{3,k}}(s)=2k$$

for s a sufficiently large integer.

For these computations, note that

(9.39) $\quad H_f(s)=1+2d(f)+e(f)$

for any finitely determined $f\in S_2$ and any $s>0$ sufficiently large.

For the other specializations, we take $a=a_1=b_1=0$ in (9.38). If $p=\min\{i; a_i\neq 0\}$, $q=\min\{j; b_j\neq 0\}$ one easily gets $f\sim(x^2+y^3, xy^p+y^q)$ and there are several possibilities:

(i) If $p\geq q$, then $f\in E^{3,q}$ for $4\leq q\leq k$ or $f\in C^3$ when $q=3$.

(ii) If $p=q-1$, then same result as above (one can use

(7.41)).

(iii) If $p<q-1$, then $f\in F^{3,p}$ for $2\le p\le k-2$ (use again (7.41)).

Since $C^3=E^{3,3}$, we infer the next result.

(9.40) COROLLARY

The singularities in $E^o_{2,2}$ *which specialize to the singularity* C^3 *are the following:* A^k *for* $k<6$; $B^{p,q}$ *for* $p+q\le 6$.

The next interesting case of specializations is the following.

(9.41) PROPOSITION

The singularities in $E^o_{2,2}$ *which specialize to the singularity* $F^{k,2}$ *are the following.*

A^m *for* $m<k+4$; $B^{p,q}$ *for* $p+q\le k+4$, $q\ge p=2,3$; $F^{m,2}$ *for* $m<k$.

PROOF

We prove only the subtle case of $B^{p,q}$-singularities. Note that $d(B^{p,q})=p-1$ and $d(F^{k,2})=2$. Hence by (9.13) one must have $p-1\le 2$ if $B^{p,q}\to F^{k,2}$. On the other hand

$$p+q=\operatorname{codim}(B^{p,q})<\operatorname{codim}(F^{k,2})=k+5 \text{ is also necessary.}$$

To show that these conditions are also sufficient, by (9.27.i) we have only to prove

$$B^{2,k+2}\to F^{k,2} \quad\text{and}\quad B^{3,k+1}\to F^{k,2}.$$

A slice to $F^{k,2}$ in the subspace of jets with $j^1=0$ is formed by the elements

(9.42) $$f=(x^2+axy+\sum_{i=2,k} a_i y^i,\, bxy+cy^2+xy^2)$$

with $a_k=1$. We can and do take more generally $a_k\neq 0$. Consider the element in this slice given by

$$g=(x^2+y^k+xy,\ xy^2)$$

One can find two convergent series $\overline{x},\overline{y}$ such that

$$\overline{x}\cdot\overline{y}=xy+x^2+y^k$$

$\bar{x}=x+y^{k-1}+\ldots,\quad \bar{y}=y+x+\ldots$

It follows that $x=\bar{x}-\bar{y}^{k-1}+\ldots$, $y=\bar{y}-\bar{x}+\ldots$ and hence

$g\sim(\bar{x}\bar{y},\ \bar{x}^3+\bar{y}^{k+1})\in B^{3,k+1}$

This shows that $B^{3,k+1}\to F^{k,2}$.

Let us come back now to the formula (9.42) and take $b=-1$, $c=1$. Consider the coordinate change $\bar{x}=x-y-xy$, $\bar{y}=y$.

Then an easy computation shows that

$$x^2=\bar{x}^2+(\sum_{i\geq 0}\bar{y}^i)^2 \bmod \bar{x}\bar{y}$$

$$xy=\sum_{i\geq 0}\bar{y}^{i+1} \bmod \bar{x}\bar{y}.$$

We choose first the parameter a such that the coefficient of $\bar{y}^{k+1}$ in x^2+axy vanishes. Then we choose a_k such that the coefficient of $\bar{y}^k$ in $x^2+axy+a_ky^k$ vanishes. This is possible since the coefficients of $\bar{y}^k$ and $\bar{y}^{k+1}$ in x^2+axy are different! Then we choose a_i for $i=1,\ldots,k-1$ such that the coefficients of all the powers of $\bar{y}$ less than the k-th are again zero.

This shows that $B^{2,k+2}\to F^{k,2}$ and ends the proof.

(9.43) REMARK

The singularity C^3 does not specialize to the singularity $F^{k,2}$ in spite of the fact that all the numerical invariants considered by us do not contradict the relation $C^3\to F^{k,2}$. Check this assertion!

The remaining cases of specializations are much simpler and we only state them below. Their proofs are good exercises for the reader.

(9.44) PROPOSITION

The singularities in $E^0_{2,2}$ *which specialize to the singularity* $F^{3,k}$ *are the following.*

A^m *for* $m<2k+3$; $B^{p,q}$ *for* $p+q\leq 2k+3$; C^3;

$E^{3,m}$ *for* $4\leq m\leq k+1$; $F^{3,m}$ *for* $2\leq m<k$.

(9.45) PROPOSITION

The singularities in $E^o_{2,2}$ *which specialize to the singularity* C^4 *are the following.*

A^m *for* $m<8$; $B^{p,q}$ *for* $p+q\leq 8$; C^3; $E^{3,4}$; $F^{m,2}$ *for* $m=3,4$.

The classification of the *unimodular* map germs $(\mathbb{C}^n,0) \to (\mathbb{C}^n,0)$, which come next in the hierarchy of singularities after the simple ones, can be found in [DG3], [DG4]. For $n=2$, a key point in this classification is the classification of pencils of binary cubic forms considered by us in Chapter 5.

Finally, concerning the classification of K-simple singularities in $E^o_{n,p}$ we remark that the case $n>p$ is also due to Giusti [Gt], while the case $n<p$ was settled by Damon [Da]. However, it seems that the specializations in the cases $n\neq p$ are not completely understood at present.

(9.46) EXERCISES

(i) Show that for $f\in A^{p+q-1}$, $g\in B^{p,q}$ the values of the corresponding Poincaré series coefficients $h_f(k)$ and $h_g(k)$ are given by the next table

k	0	1	...	p-1	p	...	q	q+1	...	p+q-1	p+q
$h_f(k)$	1	1	...	1	1	...	1	1	...	1	0
$h_g(k)$	1	2	...	2	1	...	1	0	...	0	0

Note that $f \to g$ does not imply $h_f(k)\leq h_g(k)$ for some $k\geq 0$.

(ii) Show that the correspondence $\mathrm{Grad}: m_2^2 \to E^o_{2,2}$ defined in (9.20) carries simple singularities into simple singularities. More precisely:

$$\mathrm{Grad}\ (A_k)=A^{k-1}\ , \quad \mathrm{Grad}\ (D_k)=B^{2,k-2}$$

$$\mathrm{Grad}\ (E_6)=C^3\ , \quad \mathrm{Grad}\ (E_7)=F^{3,2}\ , \quad \mathrm{Grad}\ (E_8)=C^4\ .$$

(iii) Let f^k_t be the normal form for the simple elliptic singularity $\tilde{E}_k$ (in $n\geq 3$ variables for $k=6$ and in $n\geq 2$ variables for $k=7,8$ by adding or deleting squares).

Show that codim Grad $(f_t^k)=k+4$ for $k=6,7,8$. Deduce that any map germ $f\in E^o_{2,2}$ (resp. $f\in E^o_{n,n}$ for $n\geq 3$) with codim $(f)<11$ (resp. codim $(f)<10$) is K-simple.

(iv) Let f be one of the weighted homogeneous normal forms for the simple singularities in $E^o_{2,2}$. Recall the definition of the graded vector space $N^k(f)$ used in (7.41) and take k large enough (i.e. such that $j^k f=f$). Show that $N^k(f)_s=0$ for any $s\geq 0$. Compare this with the characterization of simple function singularities by the condition $D(\underline{w},d)<d$ given after Theorem (8.26).

It seems to be an interesting open problem to find an a priori proof for the fact that the K-simple singularities $(\mathbb{C}^n,0)\to(\mathbb{C}^n,0)$ admit weighted homogeneous normal forms and to characterize these singularities by some numerical inequality as in the case of function singularities (compare with p. 145).

CHAPTER 10

CURVE AND SURFACE SINGULARITIES

§1. WEIERSTRASS NORMAL FORM AND BLOWING-UP

In this Chapter we study some special properties of complex isolated hypersurface singularities $(X,0) \subset (\mathbb{C}^n,0)$ of low dimension, namely planecurve singularities (n=2) and surface singularities (n=3).

In this first section we discuss some general facts which hold in any dimension.

Let $(X,0) \subset (\mathbb{C}^n,0)$ be a hypersurface singularity and $f=0$ be a defining equation for $(X,0)$. Then $f \in m_n \subset E_n$ and since this ring is factorial (2.15.i) one can write $f=f_1^{a_1}\ldots f_p^{a_p}$ for some prime elements $f_i \in E_n$ and some integers $a_i \geq 1$. The analytic space germ X is reduced if and only if $a_1=\ldots=a_p=1$. Moreover $X_i : f_i=0$ are precisely its irreducible components.

(10.1) LEMMA

(i) *For* n=2, (X,0) *is an isolated curve singularity if and only if* $a_1=\ldots=a_p=1$.

(ii) *For* $n\geq 3$, (X,0) *is an isolated hypersurface singularity implies that* $p=1$, $a_1=1$ (*i.e.* X *is irreducible*).

PROOF

(i) If $a_i>1$ for some i, all the partial derivatives $\frac{\partial f}{\partial x_j}$ are divisible by the factor f_i and hence any point on the irreducible component X_i is singular.

Conversely, if $a_1=\ldots=a_p=1$ then X is reduced, and for a reduced curve singularity X one clearly has dim Sing(X)<dim X. Therefore dim Sing $(X)\leq 0$.

(ii) If $p\geq 2$ it follows that any point in the intersection $X_1 \cap X_2$ is singular and hence

$\dim \operatorname{Sing}(X) \geq \dim (X_1 \cap X_2) \geq n-2 \geq 1$.

(10.2) EXERCISE (WHITNEY UMBRELLA)

Show that the surface $X: x^2-zy^2=0$ in $\mathbb{C}^3$ has the following properties

(i) $(X,0)$ is not an isolated singularity

(ii) $(X,0)$ is irreducible.

This shows that the converse of (10.1.ii) does not hold.

In fact (10.1.ii) can be strengthen to the next.

(10.3) PROPOSITION

For a normal analytic set germ $(X,0)$, *its singular set* $(SX,0)$ *satisfies* $\dim SX < \dim X-1$.

The converse implication is true when $(X,0)$ *is a complete intersection singularity.*

For the definition of normality and a proof of this result we refer to [KK] p. 315.

Next we present a famous classical result: Weiestrass Preparation Theorem, see for instance [BK], p. 338. The statement (weaker than the usual one) and the proof (quite unusual) of this result given below are in the spirit of the theory developed in our book.

(10.4) PROPOSITION (WEIERSTRASS PREPARATION THEOREM)

Let $f:(\mathbb{C}^n,0) \to (\mathbb{C},0)$ *be a finitely determined function germ and let* $s \geq 1$ *be the order of* f. *Then* f *is K-equivalent to a Weierstrass normal form* f_o *given by the formula*

$$f_o(x_1,\ldots,x_n)=x_1^s+a_2(\bar{x})x_1^{s-2}+\ldots+a_s(\bar{x})$$

where $\bar{x}=(x_2,\ldots,x_n)$ *and* $a_2,\ldots,a_s$ *are polynomials in* E_{n-1} *such that* $\operatorname{ord}(a_i) \geq i$.

Hence f_o is a monic polynomial of degree $s=\operatorname{ord}(f)$ in x_1 with coefficients in E_{n-1}.

Geometrically, the hypersurface singularity $X: f_o=0$ can be regarded as an s-sheeted branched covering space over $(\mathbb{C}^{n-1},0)$ via the projection

$$(x_1,\dots,x_n) \to (x_2,\dots,x_n) .$$

PROOF

Since s=ord (f), it follows that $j^s f$ is a nonzero homogeneous polynomial in $H^s(n,1;\mathbb{C})$. By a suitable linear change of coordinates, we can assume that $j^s f$ contains the monomial x_1^s with the coefficient 1 and no monomials $x_i x_1^{s-1}$ for $i=2,\dots,n$. In other words, $j^s f = x_1^s + \dots$, where the dots stand for monomials containing x_1 to powers $\leq s-2$. Then

$$\frac{\partial}{\partial x_1}(j^s f) = s x_1^{s-1} + \dots$$

and a complete transversal for $u=j^s f$ as in (8.11) is given by a set of monomials $x^a = x_1^{a_1} \dots x_n^{a_n}$ with $s < a_1 + \dots + a_n \leq t$ and $a_1 < s-1$.

Next we introduce some general terminology (compare with [Lm], p. 132).

(10.5) DEFINITION

Let X be a reduced analytic space and $SX \subset X$ be its singular locus. A *resolution of singularities for* X is a proper morphism $f:Y \to X$ from the nonsingular analytic space Y to X such that the restriction

$$f|Y_o : Y_o = Y \setminus f^{-1}(SX) \to X \setminus SX$$

is an analytic isomorphism and the open subset Y_o is dense in Y.

Let (X,0) be a reduced analytic space germ. A *resolution of the singularity* (X,0) is just a resolution of singularities in the above sense for a representative $\widetilde{X}$ of the germ X.

We recall now briefly the definition of the *blowing-up of a point* (*σ-process, quadratic transformation*) which is a basic tool in finding resolutions of singularities. For details look at [BK], p. 459.

Let $x_1,\dots,x_n$ be the usual coordinates on $\mathbb{C}^n$ and $(u_1:\dots:u_n)$ be the homogeneous coordinates on the complex projective space $\mathbb{P}^{n-1}$ for $n \geq 2$.

Then we consider the algebraic variety

(10.6) $\quad Z=\{(x,u)\in\mathbb{C}^n\times\mathbb{P}^{n-1};\ x_iu_j=x_ju_i \quad \text{for}\quad i,j=1,\ldots,n\}$

It is easy to see that Z is an n-dimensional complex manifold (covered by n coordinate neighbourhoods). The map $p:Z\to\mathbb{C}^n$ induced by the first projection has an exceptional fiber $E=p^{-1}(0)\simeq\mathbb{P}^{n-1}$ (called the *exceptional divisor*) and induces an isomorphism $p:Z\setminus E\to\mathbb{C}^n\setminus\{0\}$.

In other words Z (which is called the *blowing-up of* $\mathbb{C}^n$ *at the origin*) is obtained from $\mathbb{C}^n$ by replacing (or blowing-up) the point 0 with (respectively, to) the whole projective space $\mathbb{P}^{n-1}$.

Alternatively, Z can be regarded as the *disjoint* union of all the lines in $\mathbb{C}^n$ passing through the origin (the total space of the *tautological line bundle* $\mathcal{O}(-1)$-see [Lm], p. 134 for the case n=2).

Using local coordinates $x_1,\ldots,x_n$ one can define the blowing up $B_x(M)$ of any analytic space M at any smooth point $x\in M\setminus SM$. In this notation $Z=B_0(\mathbb{C}^n)$.

Assume that X:f=0 is a hypersurface singularity at the origin of $\mathbb{C}^n$ and let $p:Z\to\mathbb{C}^n$ be the blowing-up projection map.

The *total transform* X^t of X by this blowing-up is the analytic subspace of Z defined by the equation $f\circ p=0$. Note that always $E\subset X^t$ and that X^t is in general nonreduced.

The *strict (or proper) transform* X^s of X can be defined in one of the following equivalent ways:

(i) X^s is the union of the irreducible components of X^t different from E (with reduced structure).

(ii) X^s is the closure in Z of $p^{-1}(X\setminus 0)$.

It is usual to say that the proper transform X^s is obtained by blowing up the point $0\in X$ (compare to [Ha], p. 165).

(10.7) EXAMPLE

Let $X:f=x_1^d+\ldots+x_n^d=0$ be a homogeneous isolated singularity. On the open subset $Z_1=\{(x,u)\in Z;\ u_1\neq 0\}$ we can take as coordinates the functions $x_1,u_2,\ldots,u_n$. Then the total transform X^t is given in Z_1 by the equation

$$x_1^d(1+u_2^d+\ldots+u_n^d)=0$$

There are two irreducible components in X^t: the exceptional divisor E corresponding to the equation $x_1=0$ in Z_1 and the strict transform X^s corresponding to the equation $1+u_2^d+\ldots+u_n^d=0$ in Z_1.

One usually writes formally (i.e. in the language of divisors cf. [Ha])

$$X^t=X^s+d\cdot E$$

and d is called the *multiplicity* of E in the total transform X^t. Note that d=mult (X,0).

Moreover, it is easy to check that X^s is *smooth* and the restriction of p induces a *resolution of singularities* $q:X^s\to X$.

We also remark that $q^{-1}(0)=X^s\cap E$ is precisely the Fermat hypersurface $H_d:u_1^d+\ldots+u_n^d=0$ in $\mathbb{P}^{n-1}\simeq E$ and one can regard X^s as the total space of the tautological bundle $\mathcal{O}(-1)$ restricted to H_d.

In the next sections we shall use the blowing-ups to find resolutions for all plane curve singularities and for the surface singularities in $\mathbb{C}^3$ corresponding to the simple singularities A-D-E from Chapter 8.

§2. INVARIANTS OF PLANE CURVE SINGULARITIES

We start the discussion of plane curve isolated singularities with the next algebraic result.

(10.8) PROPOSITION (HENSEL'S LEMMA)

Let $f\in E_2$ *be an analytic map germ such that*

$$j^sf=(a_1x+b_1y)^{k_1}\ldots(a_px+b_py)^{k_p}$$

is the factorization into distinct linear factors of its s-*jet, where* $s=\text{ord}\,(f)=k_1+\ldots+k_p$.

Then there is a factorization of the element f *in* E_2 *of the form* $f=f_1\ldots f_p$ *such that* ord $(f_i)=k_i$ *and* $j^{k_i}f_i=(a_ix+b_iy)^{k_i}$ *for* $i=1,\ldots,p$.

Note that we do not claim (and it is not true in general!) that the elements f_i above are irreducible.

PROOF

It is clearly enough to show the following:

If $j^sf=P\cdot Q$ where P and Q are homogeneous polynomials of degree k and ℓ respectively in x,y and without common factors, then f admits a factorization $f=g\cdot h$ in E_2 such that $j^kg=P$ and $j^\ell h=Q$.

To prove this we write

$$f=f_s+f_{s+1}+\ldots,\quad g=g_k+g_{k+1}+\ldots,\quad h=h_\ell+h_{\ell+1}+\ldots$$

where f_m , g_m , h_m stand for the homogeneous components of degree m in the Taylor expansions of f,g and h respectively.

We intend to solve inductively the sequence of equations

$$(10.9)\qquad f_m=g_k\cdot h_{m-k}+g_{k+1}\cdot h_{m-k-1}+\ldots+g_{m-\ell}\cdot h_\ell \quad \text{for } m\geq s.$$

To start with, we put $g_k=P$, $h_\ell=Q$. Assume next that $g_k,\ldots,g_{k+n-1}$, $h_\ell,\ldots h_{\ell+n-1}$ have already been determined. To find g_{k+n} , $h_{\ell+n}$ we use the equation (10.9) which we put in the form

$$(10.10)\qquad P\cdot h_{\ell+n}+Q\cdot g_{k+n}=R$$

with R a known polynomial in H^{s+n}, where we write for simplicity $H^t=H^t(2,1;\mathbb{C})$.

Consider the linear map

$$T:H^{\ell+n}\times H^{k+n}\to H^{s+n},\quad T(A,B)=PA+QB\ .$$

Since G.C.D. (P,Q)=1, it follows that

$$\ker T=\{(QC,\ -PC);\ C\in H^n\}\ .$$

In particular, dim (ker T)=dim $H^n=n+1$.

Since dim $(H^{\ell+n}\times H^{k+n})=s+2n+2$ and dim $H^{s+n}=s+n+1$, a well-

-known result from Linear Algebra yields im $T=H^{s+n}$. This shows that the equation (10.10) has a (non unique) solution.

Proceeding in this way, we get a factorization $f=g\cdot h$ in the ring $\mathbb{C}[[x,y]]$ of formal power series. An application of Artin's Approximation Theorem (6.14) shows that, for any integer $N>0$, there exist $\bar{g},\bar{h}\in E_2$ such that: $f=\bar{g}\cdot\bar{h}$, $j^N\bar{g}=j^N g$ and $j^N\bar{h}=j^N h$.

(10.11) REMARK

In order to avoid the use of Artin's Theorem, one can define an *algebroid plane curve singularity* to be just a principal reduced ideal in $\mathbb{C}[[x,y]]$ and part of the theory can be developed for these objects.

Let $X:f=0$ be an isolated curve singularity at the origin of $\mathbb{C}^2$. Recall that mult $(X)=$ord (f) and the multiplicity mult (X) is one if and only if X is smooth.

(10.12) EXERCISE

Show that mult $(X)=2$ if and only if X is a simple singularity of type A_k for some $k\geq 1$. [The last assertion means that $f\in A_k$. Check that this property does not depend on the choice of the defining equation for X, but only on the isomorphism class of X. Hint: use (3.16).]

Let now $s=$mult (X) and consider the decomposition $f=f_1\cdots f_p$ of f described in Hensel's Lemma (10.8). The isolated curve singularities $X_i:f_i=0$ are called the *tangential components* of the curve singularity X.

The line $L_i:a_ix+b_iy=0$ is called the *tangent line* to the component X_i.

In order to understand these notions well, we recall the next definitions (for more details [BK], p. 440).

(10.13) DEFINITION

Let $X:f=0$, $Y:g=0$ be two curve singularities at the origin $0\in\mathbb{C}^2$. Then the (*local*) *intersection number* of X and Y is the positive integer or ∞

$$(X,Y)=(X,Y)_0=\dim E_2/(f,g)\leq\infty\ .$$

Note that this dimension is finite if and only if the curve singularities X and Y have no common irreducible components.

A line $L:\ell=0$ is said to be *tangent to the curve singularity* $X:f=0$ if $(X,L)>\text{mult}\,(X)$.

With these definitions at hand, it is clear that the tangential component X_i is precisely the union of the *branches* (i.e. irreducible components) of X for which the line L_i is a tangent.

In particular, a branch of X has a unique tangent line, but not conversely.

(10.14) EXAMPLE

Consider the simple curve singularity

$$A_k:x^{k+1}+y^2=0 \quad \text{for} \quad k\geq 1.$$

This singularity is irreducible for k even and has two irreducible components for k=2m-1 odd:

$$x^{2m}+y^2=(x^m+iy)(x^m-iy) \quad , \quad \text{where} \quad i=\sqrt{-1} \quad .$$

In the latter case both branches are smooth.

Now, for k=1, there are two tangential components and the tangent lines $x\pm iy=0$ coincide in this case with the branches of the singularity A_1.

For $k\geq 2$, there is only one tangential component with tangent line y=0.

It is more usual to write the equivalent normal form $A_k:x^{k+1}-y^2=0$ which allows one to draw the next picture (in the real plane).

The classical terminology for the A_k-curve singularities is the next: A_1=node, A_2=cusp, A_3=tacnode, A_4=ramphoid cusp,...

We show next that the Boardman symbol can be used to define an interesting invariant for any plane curve isolated singularity X:f=0.

Since f is finitely $\mathcal{K}$-determined by (6.39) we can apply (10.4) and take f a Weierstrass normal form, namely

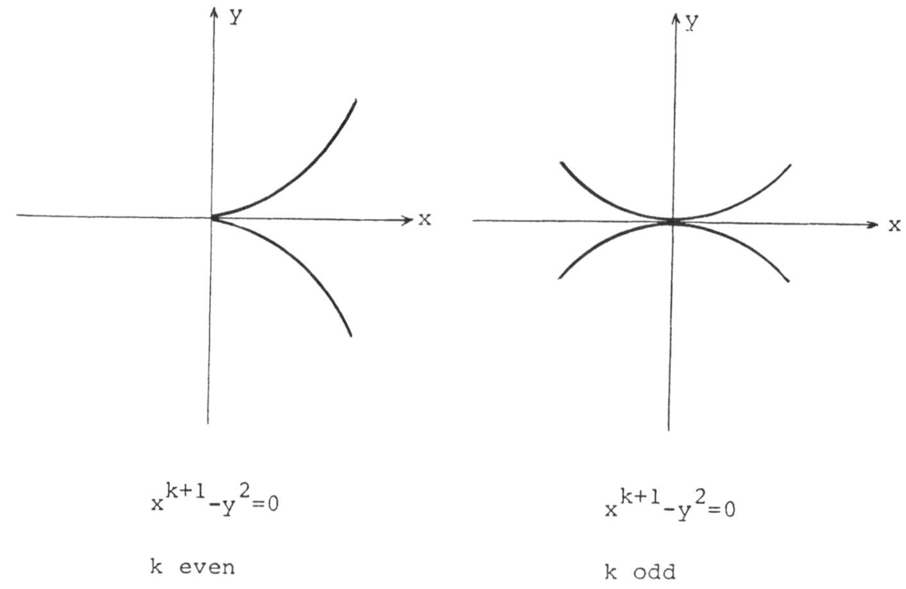

$x^{k+1}-y^2=0$

k even

$x^{k+1}-y^2=0$

k odd

(10.15) FIGURE

$$f=x^s+a_2(y)x^{s-2}+\ldots+a_s(y), \quad \text{where} \quad s=\text{mult}(X).$$

The coefficients $a_j(y)$ or even the numbers ord $(a_j)\geq j$ are not invariants of the singularity X. Nevertheless, one has the next result.

(10.16) PROPOSITION

The Boardman symbol of the function germ f *above is*

$$(\underbrace{2,2,\ldots,2}_{(s-1)\text{-times}},\ \underbrace{1,1,\ldots,1}_{k\text{-times}},\ 0,0,\ldots)$$

where $k=\min\limits_{2\leq j\leq s}(\text{ord}\,(a_j)-j)$.

In particular, it follows that k is a K-invariant of the function germ f by (9.6.i) and hence we can write $k=k(X)$.

PROOF

Let I be the principal ideal (f) in E_2. Note that the first (s-1) Jacobian critical extensions of this ideal are all of the type $\Delta^2(\Delta^2(\ldots(\Delta^2(I))))$ and the last of these ideals (call it J) is generated by f and all its partial derivatives up to and including order s-1.

Since $\frac{\partial^{s-1} f}{\partial x^{s-1}} = x$, it is easy to see that in fact $J=(x,y^{k+1})$ where $k=\min(\operatorname{ord}(a_j)-j)$ as in the statement. And now it is an easy matter to finish the proof.

In the following, we establish a basic formula relating the Milnor number of a plane curve singularity with some others of its invariants.

Before giving the statement, we recall some useful facts on the famous *Puiseux parametrizations*. As usual, for details we send to [BK].

For any irreducible plane curve singularity $(X,0) \subset (\mathbb{C}^2,0)$ there is an analytic map germ $c:(\mathbb{C},0) \to (X,0)$ such that

(i) $c(D) \subset \tilde{X}$, $c^{-1}(0)=0$;

(ii) c induces an analytic isomorphism $D\setminus\{0\} \to \tilde{X}\setminus\{0\}$, where D is a small disc centered at the origin in $\mathbb{C}$ and $\tilde{X}$ is a suitable representative of the germ (X,0).

The map c above is called the Puiseux parametrization of the branch (X,0) and c is in fact nothing else but the normalization (resolution) of the singularity (X,0).

Choosing suitable coordinates in $\mathbb{C}^2$, the Puiseux parametrization c is given by the formulas:

$$x=t^n,\ y=\sum_{i\geq m} a_i t^i \text{ for some } m>n=\operatorname{mult}(X).$$

In more classical notation, one writes

$$y=\sum_{i\geq m} a_i x^{i/n}$$

and hence y is a fractional power series in x.

In the sequel we need only the next property of the

Puiseux parametrization [BK], p. 411: for any plane curve singularity $Y:g=0$ at the origin of $\mathbb{C}^n$ the intersection number (X,Y) equals ord $(g\circ c)$. Now if $X:f=0$ is any plane curve singularity, one can write formally $X=a_1X_1+\ldots+a_pX_p$ where $f=f_1^{a_1}\ldots$ $\ldots f_p^{a_p}$ is the prime decomposition of f in E_2 and X_i: $f_i=0$ are the (reduced) irreducible components of X. Since the intersection number is linear:

$$(X,Y)=\sum_{i=1,p} a_i(X_i,Y)$$

it follows that one can use Puiseux parametrizations to compute (X,Y) in general.

Consider a Weierstrass type equation for an isolated plane curve singularity

$$X:f=x^s+a_2(y)x^{s-2}+\ldots+a_s(y)=0$$

where we do *not* assume that $s=\text{mult}\,(X)$, ord $(a_j)\geq j$.

Then f is a polynomial in $\mathbb{C}\{y\}[x]$ and one can define its *discriminant* $D_y(f)\in\mathbb{C}\{y\}$, which is nothing else but the *resultant* of f and $\frac{\partial f}{\partial x}$.

We consider the positive integer

$$\delta_y(X)=\delta_y(f)=\text{ord}\,D_y(f).$$

(10.17) PROPOSITION

With the above notations

$$\delta_y(X)=\mu(X)+(L,X)-1$$

where $\mu(X)=\mu(f)$ *is the Milnor number of the singularity* $(X,0)$ *and* L *denotes the coordinate axis* $y=0$.

PROOF

Let Y be the plane curve singularity defined by the equation $\frac{\partial f}{\partial x}=0$ at the origin of $\mathbb{C}^2$. Then the classical definition of the intersection number (see [BK], p. 231) implies

$$(X,Y)=\text{ord}\,D_y(f)$$

Let $Y=a_1Y_1+\ldots+a_pY_p$ be the decomposition of Y according to its irreducible components Y_i. Then

$$(X,Y)=\sum_{i=1,p} a_i(X,Y_i) .$$

Let c be a Puiseux parametrization for the branch Y_i. Then we have $(X,Y_i)=\text{ord }(f\circ c)$ and by the formula

$$\frac{d}{dt}(f\circ c)=(\frac{\partial f}{\partial x}\circ c)\cdot\frac{dc^1}{dt}+(\frac{\partial f}{\partial y}\circ c)\cdot\frac{dc^2}{dt}$$

we get $\text{ord }(f\circ c)=\text{ord}(\frac{\partial f}{\partial y}\circ c)+\text{ord }(y\circ c)$. Here c^1,c^2 are the components of c and note that $\frac{\partial f}{\partial x}\circ c=0$.

In other words, we obtain

$$(X,Y_i)=(Z,Y_i)+(L,Y_i), \quad \text{where} \quad Z:\frac{\partial f}{\partial y}=0$$

Multiplying by a_i and making the sum give

$$(X,Y)=(Z,Y)+(L,Y).$$

By the definitions of the Milnor number and the intersection number one has $(Z,Y)=\mu(X)$, thus ending the proof. This proof is due essentially to Teissier, see [Te], p. 317.

§3. BLOWING-UP PLANE CURVE SINGULARITIES

In this section we describe in detail how a plane curve singularity $(X,0)$ and its invariants mult (X), $k(X)$ and $\mu(X)$ change under a blowing-up. And we end our discussion of curve singularities by giving a characterization of simple plane curve singularities A-D-E in terms of their reduced total transform under a blowing-up.

Let $X:f=0$ be an isolated curve singularity at the origin of $\mathbb{C}^2$. Then f is defined on some open set U containing the origin and we identify the germ $(X,0)$ with the representative $X=\{x\in U;\ f(x)=0\}$.

Let $V=B_o(U)$ denote the blowing up of U at the origin and

$p:V \to U$ denote the projection. In the notations of the first section in this Chapter, $E=P^1$ is covered by 2 affine coordinate charts: $\{(u_1:u_2);u_i\neq 0\}$ for $i=1,2$. The corresponding open sets V_1 and V_2 in V admit coordinates (x,u_2) and (y,u_1) respectively such that the projection p is given by

(10.18) $\quad (x,u_2) \to (x,xu_2)$ on V_1; $\quad (y,u_1) \to (yu_1,y)$ on V_2 .

This implies that $x=0$ on V_1 and $y=0$ on V_2 are the equations for the exceptional divisor $E\subset V$. Assume that

(10.19) $\quad f=f_s(x,y)+f_{s+1}(x,y)+\dots$

is the Taylor expansion of f, with $s=\text{mult}\ (X)$ and f_m the homogeneous components of this expansion. Then $f_m(x,xu_2)=$ $=x^m\ f_m(1,u_2)$, $f_m(yu_1,y)=y^m f_m(u_1,1)$ show that the exceptional divisor E is contained in the total transform X^t with multiplicity $s=\text{mult}\ (X)$. In other words

$X^t=X^s+s\cdot E$ as in Example (10.7).

Next we investigate the singularities of the proper transform $\bar{X}=X^s$. Since p induces an isomorphism $V\setminus E\simeq U\setminus\{0\}$ and 0 is the only singularity of X (if we choose U small enough), it follows that the singularities of the proper transform $\bar{X}$ are contained in the intersection $\bar{X}\cap E$.

(10.20) LEMMA

A point $(u_1:u_2)\in E$ *is in the intersection* $\bar{X}\cap E$ *if and only if the line* $L:u_2x-u_1y=0$ *is a tangent line to the curve singularity* $(X,0)$.

PROOF

Assume that $a=(u_1:u_2)$ has $u_1\neq 0$. Then we can write $a=(1:u)$ with $uu_1=u_2$. Using (10.19), the equation of $\bar{X}$ in V_1 is the following

(10.21) $\quad f_s(1,u)+xf_{s+1}(1,u)+\dots=0$

A point in V_1 with coordinates (x,u) is in the intersection $\bar{X}\cap E$ if and only if $x=f_s(1,u)=0$. This means that the polynomial $f_s(x,y)$ is divisible by the linear form y-ux.

This ends the proof, since the situation in the open set V_2 is symmetric.

(10.22) COROLLARY

The connected components of the proper transform $\overline{X}$ *are in bijection with the tangential components of the singularity* (X,0) (*if we choose* U *small enough*).

Indeed, with the notation from (10.8), there are exactly p lines $L_i: a_i x + b_i y = 0$ $(i=1,\dots,p)$ tangent to the curve singularity (X,0). Let $\overline{x}_1,\dots,\overline{x}_p$ be the corresponding points in $\overline{X}\cap E$. One can say that *by blowing up the singularity* (X,0) *has split into the singularities* $(\overline{X},\overline{x}_i)$ *for* $i=1,\dots,p$.

Then (10.21) implies that mult $(\overline{X},\overline{x}_i)$ is less or equal to the multiplicity of u as a root of the equation $f_s(1,u)=0$, where $\overline{x}_i=(1:u)$. This gives us the next

(10.23) COROLLARY

$$\text{mult }(X,0) \geq \sum_{i=1,p} \text{mult }(\overline{X},\overline{x}_i) .$$

Since mult $(\overline{X},\overline{x}_i)\geq 1$ for any i, it follows that mult $(X,0)>$ $>$mult $(\overline{X},\overline{x}_j)$ for all $j=1,\dots,p$ with the possible exception when p=1. Let us consider this possibility closer.

Since p=1, in suitable coordinates we can write $X: f=x^s+$ $+a_2(y)x^{s-2}+\dots+a_s(y)=0$ with s=mult (X), ord $(a_i)>i$ for all $i=2,\dots,s$.

The tangent line x=0 corresponds to the point $\overline{x}=(0:1)\in$ $\in \overline{X}\cap E$ and hence we have to work in the coordinate chart V_2.

A local equation for $\overline{X}$ around $\overline{x}$ is given by $u^s+\overline{a}_2(y)u^{s-2}+\dots+\overline{a}_s(y)=0$, with $\overline{a}_i(y)=a_i(y)\cdot y^{-i}$ for $i=2,\dots,s$.

Now there are two possibilities:

(i) mult $(\overline{X},\overline{x})<$mult (X,0), when there is some j such that ord $(\overline{a}_j)<j$.

(ii) mult $(\overline{X},\overline{x})=$mult (X,0), when ord $(\overline{a}_j)\geq j$ for all $j=2,\dots,s$. But then we clearly have $k(\overline{X},\overline{x})<k(X,0)$ where the invariant k(X)=k(X,0) is that introduced in (10.16).

(10.24) COROLLARY

For a curve singularity $(X,0)$ *with an unique tangential component either* mult $(\overline{X},\overline{x})<$mult $(X,0)$ *or* mult $(\overline{X},\overline{x})=$mult $(X,0)$ *but then* $k(\overline{X},\overline{x})<k(X,0)$.

The last two corollaries show that *any singularity* $(\overline{X},\overline{x}_i)$ *of the proper transform* $\overline{X}$ of a curve singularity $(X,0)$ *is strictly less singular than the original singularity* $(X,0)$.

Next we want to reobtain this result by using a relation among the Milnor numbers $\mu(X,0)$ and $\mu(\overline{X},\overline{x}_i)$ for $i=1,\ldots,p$. Before giving the statement, we recall some facts about the *intersection number* (C_1,C_2) of two curves C_1, C_2 on a smooth complex surface such that $C_1\cap C_2$ consists of finitely many points $x_1,\ldots,x_n$. By definition

(10.25) $$(C_1,C_2)=\sum_{i=1,n}(C_1,C_2)_{x_i}$$

where the local intersection numbers $(C_1,C_2)_{x_i}$ were introduced in (10.13).

We need the following property.

(10.26) LEMMA

Let $X:f=0$ *and* $Y:g=0$ *be two curves in some open set* U *containing the origin in* $\mathbb{C}^2$ *such that* $X\cap Y=\{0\}$. *Let* $\overline{X}$, $\overline{Y}$ *be the proper transforms of* X, Y *by blowing-up the origin. Then*

$$(X,Y)_o=(\overline{X},\overline{Y})+\text{mult }(X,0)\cdot\text{mult }(Y,0)\ .$$

A proof for this can be found in [BPV], p. 66.

The result we are looking for is the next.

(10.27) PROPOSITION

$$\mu(X,0)-1=s(s-1)+\sum_{i=1,p}(\mu(\overline{X},\overline{x}_i)-1)$$

where $s=$mult $(X,0)$ *and* p *is the number of tangential components of* $(X,0)$.

PROOF

Let X be given in $U\subset\mathbb{C}^2$ by a Weiestrass polynomial

$f=x^s+a_2(y)x^{s-2}+\ldots+a_s(y)=0$, ord $(a_i)\geq i$. Then $y=0$ is not a tangent line to $(X,0)$ and hence $\overline{X}\cap E$ is contained in the coordinate chart V_2. Note that $p:V_2\simeq\mathbb{C}^2\to\mathbb{C}^2$ is given by $p(u,y)=(uy,y)$ and we let $\overline{g}$ denote the *proper transform of a function* $g:(\mathbb{C}^2,0)\to\mathbb{C}$ i.e. we take first $g\circ p$ and then divide out the highest power of y.

It is an easy remark that

$$(10.28)\qquad \overline{\frac{\partial f}{\partial x}}=\frac{\partial}{\partial u}(\overline{f}) \ .$$

We take $Y:\frac{\partial f}{\partial x}=0$ as in (10.17). By (10.26) we get

$$\delta_y(X,0)=(X,Y)_o=s(s-1)+(\overline{X},\overline{Y}) \ .$$

When U is small enough, all the intersection points in $\overline{X}\cap\overline{Y}$ are in fact contained in E (since $X\cap Y\cap U=\{0\}$, a good exercise for the reader!).

For any point $\overline{x}\in\overline{X}\cap\overline{Y}\cap E$, the formula (10.28) yields

$$(\overline{X},\overline{Y})_{\overline{x}}=\delta_y(\overline{X},\overline{x}) \ .$$

Using Proposition (10.17) and Lemma (10.26) we get

$$\mu(X,0)+s-1=\delta_y(X,0)=(X,Y)_o=s(s-1)+\sum_{i=1,p}(\overline{X},\overline{Y})_{\overline{x}_i}=$$

$$=s(s-1)+\sum_{i=1,p}\delta_y(\overline{X},\overline{x}_i)=s(s-1)+\sum_{i=1,p}(\mu(\overline{X},\overline{x}_i)+s_i-1)$$

where $s_i=(E,\overline{X})_{\overline{x}_i}$. Since $s=\sum_{i=1,p}s_i=(E,\overline{X})$ by the formula (10.21), this ends the proof.

The above proof is due to Pham [Ph].

(10.29) COROLLARY

With the above notations, if $s=\text{mult}\ (X)\geq 2$, *then*

$\mu(X,0)>\mu(\overline{X},\overline{x}_i)$ *for any* $i=1,\ldots,p$.

PROOF

Since $p\leq s$, one has

$$\mu(X,0)-1=\mu(\bar{X},\bar{x}_i)-1+s(s-1)+\sum_{j\neq i}(\mu(\bar{X},\bar{x}_j)-1)\geq$$

$$\geq\mu(\bar{X},\bar{x}_i)-1+s(s-1)-p+1\geq\mu(\bar{X},\bar{x}_i)-1+(s-1)^2>\mu(\bar{X},\bar{x}_i)-1 \quad .$$

One can use Corollaries (10.23), (10.24) or Corollary (10.29) to give a proof of the next fundamental result (compare to [BK] p. 496).

(10.30) THEOREM

The singularities of any curve on a complex surface can be resolved by a locally finite sequence of blowing-ups.

Finally in this section, we want to analyse the behaviour of the simple plane curve singularities under blowing-up

$$A_k : x^{k+1}+y^2=0 \ , \ k\geq 1; \quad D_k : x^2y+y^{k-1}=0 \ , \quad k\geq 4 \ ;$$

$$E_6 : x^3+y^4=0; \quad E_7 : x^3+xy^3=0; \quad E_8 : x^3+y^5=0 \ .$$

For convenience of notation, we let A_o denote a smooth branch $x+y^2=0$.

The next table describes the singularities $(\bar{X},\bar{x}_i)$ of the proper transform $\bar{X}$ of a simple curve singularity $(X,0)$ as above and also the singularities of the *reduced total transform* $\bar{X}\cup E=X^t_{red}$, at the same points $\bar{x}_i$.

(10.31) TABLE

X	A_1	A_2	A_k, $k\geq 3$	D_4	D_5	D_k, $k\geq 6$	E_6	E_7	E_8
$\bar{X}$	$2A_o$	A_o	A_{k-2}	$3A_o$	$2A_o$	A_oA_{k-5}	A_o	A_1	A_2
$\bar{X}\cup E$	$2A_1$	A_3	D_{k+1}	$3A_1$	A_1A_3	A_1D_{k-2}	A_5	D_6	E_7

We prove a single case for illustration, namely $X=D_5$ and leave to the reader the useful task of checking the other entries in this Table.

So let $X:x^2y+y^4=0$ and recall that $V=B_o(\mathbb{C}^2)=V_1 \cup V_2$, with local coordinates (x,u) on V_1 such that $y=ux$.

Then $\bar{X} \cap V_1$ is given by the equation $u-u^4x=0$. Now $u=0$ defines a line L which intersects the exceptional divisor $E:x=0$, while the curve $P:1-u^3x=0$ clearly does not intersect E.

In the second coordinate chart V_2 we have coordinates (v,y) with $x=vy$. Hence $\bar{X} \cap V_2$ is given by the equation $v^2-y=0$. This represents again the "parabola" P but this time in V_2 and intersecting nontransversally E.

All this can be represented by picture (10.32).

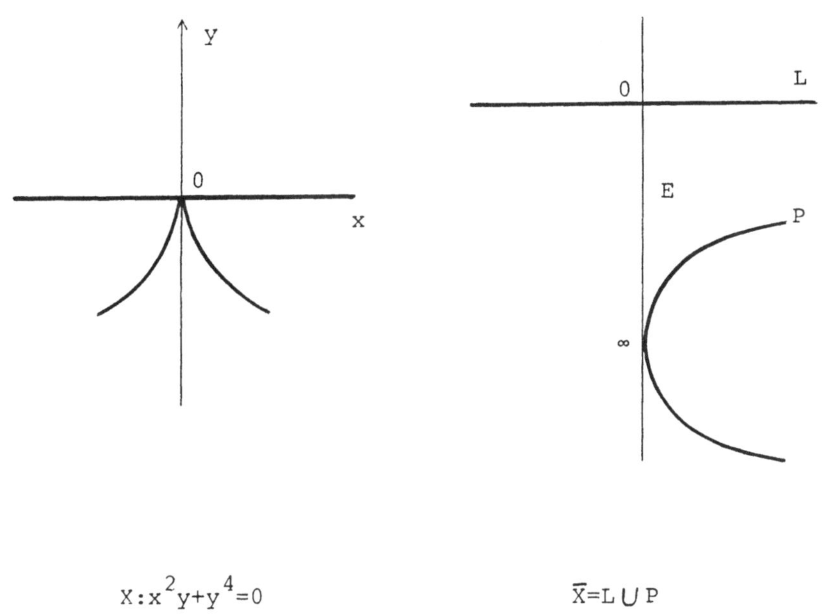

(10.32) FIGURE

Hence the proper transform $\bar{X}=L \cup P$ is smooth at the points $\bar{x}_1=0=(1:0)$ and $\bar{x}_2=\infty=(0:1)$. It is also clear that the reduced total transform $X^t_{red}=L \cup P \cup E$ has an A_1-singularity at the point $\bar{x}_1$ and an A_3-singularity at $\bar{x}_2$.

We can introduce the next definition (compare to [BPV], p. 61).

(10.33) DEFINITION

An isolated plane curve singularity $(X,0)$ is *blowing-up simple* if mult $(X,0)\leq 3$ and mult $(\bar{X}\cup E,\bar{x}_i)\leq 3$ for all the points $\bar{x}_i\in \bar{X}\cap E$.

And one has the following interesting result

(10.34) PROPOSITION

A plane curve singularity is blowing-up simple if and only if it is a simple singularity (i.e. of type A, D *or* E).

PROOF

Table (10.31) gives obviously the implication: $(X,0)$ simple $\Longrightarrow$ $(X,0)$ blowing-up simple.

Conversely, assume now that $(X,0)$ is a blowing-up simple singularity.

If mult $(X,0)\leq 2$, then $(X.0)$ is a singularity of type A_k for some $k\geq 0$ and we are ready.

If mult $(X,0)=3$ and $(X,0)$ has at least two tangential components, then $(X,0)$ is of type D_k for some $k\geq 4$.

The remaining case is when mult $(X,0)=3$ and there is exactly one tangential component.

Then we can write a Weierstrass normal form equation for X:

$$f=x^3+a_2(y)x+a_3(y)=0 \quad \text{with} \quad \text{ord}\ (a_i)>i \quad \text{for} \quad i=2,3.$$

In the coordinate chart V_2 , the exceptional divisor E is given by $y=0$ and the proper transform $\bar{X}$ by

$$\bar{f}=u^3+\bar{a}_2(y)u+\bar{a}_3(y)=0 \quad \text{with} \quad \bar{a}_i(y)=a_i(y)\cdot y^{-i} \quad \text{for} \quad i=2,\ 3.$$

Since $(X,0)$ is blowing-up simple, it follows that ord $(\bar{f})\leq 2$ and this is equivalent to

$$\text{either ord}\ (a_2)\leq 3 \quad \text{or} \quad \text{ord}\ (a_3)\leq 5.$$

If we go back to the proof of Proposition (8.21) we see that in these conditions $(X,0)$ is isomorphic to a singularity of type E_6 , E_7 or E_8.

Other interesting connections between blowing-up plane curve singularities and classification of function singularities can be found in a paper by Siersma [Si].

§4. BASIC FACTS ON RESOLUTION OF SURFACE SINGULARITIES

In the remaining part of this Chapter we consider isolated surface singularities $X:f=0$, defined by an analytic function germ f at the origin of $\mathbb{C}^3$.

For such a singularity (or, more generally, for a normal surface singularity), let $p:M \to X$ be a resolution of the singularity $(X,0)$ as in (10.5).

Then the *exceptional set* $E=p^{-1}(0)$ is compact (p proper), one-dimensional ($M_o=M\setminus E$ is dense in M) and connected (otherwise $M_o \simeq \widetilde{X}\setminus\{0\}$ is not connected for a representative $\widetilde{X}$ small enough for $(X,0)$-contradicting the irreducibility of X (10.1)).

Hence $E=E_1\cup\ldots\cup E_s$, where each E_i is a compact irreducible curve in M. Moreover it is possible to arrange that the curves E_i are smooth, the intersections $E_i\cap E_j$ for $i\neq j$ are transverse and $E_i\cap E_j\cap E_k=\emptyset$ for $i\neq j\neq k$.

This follows essentially from the embedded resolution of singularities for curves by blowing-ups as in Theorem (10.30).

A resolution with the above properties is called *good*. If in addition it satisfies $(E_i,E_j)\leq 1$ for any $i\neq j$, then it is called a *very good* resolution.

If $p:M \to X$ is a good resolution as above, then the matrix

(10.35) $\quad ((E_i,E_j))_{i,j=1,\ldots,s}$

is called the *intersection matrix* of the resolution and it is known that this matrix is negative definite [La], p. 49. Here the intersection number (E_i,E_j) for $i\neq j$ is defined as in (10.25) and under the transversality assumption on $E_i\cap E_j$ this number equals the number of points in this intersection. For the definition of the *self intersection number* (E_i,E_i) we refer to [La], p. 14.

Conversely, given a collection $E=E_1 \cup \ldots \cup E_s$ of smooth compact curves in a two-dimensional manifold M, intersecting transversely and such that the intersection matrix $((E_i,E_j))_{i,j=1,\ldots,s}$ is negative definite, the quotient space M/E obtained from M by *contracting* (*blowing-down*) the set E to a point $0 \in M/E$ has a normal complex structure such that the natural projection $q:M \to M/E$ is analytic [La], p. 60.

(10.36) EXAMPLE

An *exceptional curve of the first kind* is a rational curve C (i.e. C is smooth and with genus $g(C)=0$ or, equivalently $C \simeq \mathbb{P}^1$) on a smooth surface M such that $(C,C)=-1$.

Such a curve arises when blowing-up a smooth point on a surface. Indeed, $B_o(\mathbb{C}^2)$ can be identified to the total space of the line bundle $\mathcal{O}(-1)$ on $\mathbb{P}^1$ in such a way that the exceptional divisor E corresponds to the zero section. This implies that $(E,E)=-1$ by [La], p. 15. Since clearly $E \simeq \mathbb{P}^1$, this shows that E is indeed an exceptional curve of the first kind.

Conversely, any exceptional curve of the first kind blows-down to a smooth point [La], p. 80.

This example justifies the next.

(10.37) DEFINITION

A resolution $p:M \to X$ is called *minimal* if no one of the components E_i of the exceptional set $E=p^{-1}(0)$ is an exceptional curve of the first kind.

A minimal resolution $p:M \to X$ has the following universal property: given any other resolution $p':M' \to X$, there is a unique map $q:M' \to M$ such that $p'=p \circ q$. In particular, a minimal resolution is unique up to isomorphism. And for any normal singularity $(X,0)$ there is a minimal resolution $p:M \to X$ which is good [La], p. 91.

When dealing with a very good resolution, it is convenient to associate a *weighted graph* to the intersection matrix. The vertices of the graph are $v_1,\ldots,v_s$ in bijection with the components $E_1,\ldots,E_s$ of the exceptional set E. Two different vertices v_i and v_j are joined if and only if $(E_i,E_j)=1$. And finally the vertex E_i is weighted by the integer (E_i, E_i) for all $i=1,\dots,s$.

The weighted graph obtained in this way is called the *dual graph of the resolution*. Here "dual" is used since the point v_i corresponds to the curve E_i (usually a line $\mathbb{P}^1$) and the line joining v_i to v_j corresponds to the intersection point $E_i \cap E_j$.

Starting with a normal surface singularity $(X,0)$, there are two basic methods to get a resolution:

(i) alternately blow up points and normalize;

(ii) blow up points and smooth curves.

(10.38) DEFINITION

A normal surface singularity $(X,0)$ is called *absolutely isolated* if it may be resolved by blowing up points only.

(10.39) EXAMPLE

Let X be the isolated singularity in $\mathbb{C}^3$ given by

$$X: f = x^d + y^d + z^d = 0 \, , \quad \text{for some} \quad d \geq 2.$$

Then the computations in Example (10.7) show that the first strict transform $\overline{X} = X^s$ is already smooth and hence provides us with a (minimal!) resolution for $(X,0)$.

It follows that $(X,0)$ is an absolutely isolated singularity, for any $d \geq 2$. Moreover, the Example (10.7) shows that the exceptional set E in this case is identified to the zero section of the total space of the line bundle $L = \mathcal{O}(-1)|_C$, where the curve C is given by

$$C = \{u \in \mathbb{P}^2; \quad u_1^d + u_2^d + u_3^d = 0\}$$

It follows that $(C,C) = \deg(L) = -d$ as in [La], p. 14.

Therefore the dual graph of this resolution is

$$\underset{-d}{\bullet}$$

and the exceptional set $E \simeq C$ is a smooth curve of genus $g(C) = \dfrac{(d-1)(d-2)}{2}$.

Recall that for a projective algebraic smooth variety X one defines its *geometric genus*

$$p_g(X)=\dim_{\mathbb{C}} H^0(X, \Omega_X^n)$$

where n=dim X, Ω_X^n denotes the line bundle of holomorphic n-forms on X and $H^0(X,\Omega_X^n)$ denotes the (finite dimensional!) vector space of global sections of this line bundle (see for instance [Ha], p. 181).

When X is a smooth curve, $p_g(X)=g(X)$ the *topological genus* of X and the curve X is called *rational* if $p_g(X)=0$.

Assume now that (X,0) is a normal surface singularity and let $X'=X\setminus\{0\}$ be the smooth part of a small enough representative for (X,0). Then the vector space $H^0(X',\Omega_{X'}^2)$ is infinite dimensional, but we can get a finite dimensional vector space in the following way.

Let $p:M \to X$ be the minimal resolution for the singularity (X,0). Note that since p induces an isomorphism $M\setminus p^{-1}(0) \to X'$, there is a well-defined injective linear map

$$r:H^0(M,\Omega_M^2) \to H^0(X',\Omega_{X'}^2)$$

given by restriction of sections.

Then it is known that

$$\operatorname{coker}(r)=H^0(X',\Omega_{X'}^2)/r(H^0(M,\Omega_M^2))$$

is a finite dimensional vector space [P] and this makes possible the next.

(10.40) DEFINITION

The *geometric genus* of a normal surface singularity (X,0) is the positive integer

$$p_g(X,0)=\dim(\operatorname{coker}(r)) .$$

A normal surface singularity (X,0) is called *rational* if $p_g(X,0)=0$.

Note that the condition $p_g(X,0)=0$ can be interpreted also in the following way: *for any 2-form ω on X' the pull-back*

$p^*(\omega)$ *can be extended holomorphically to* M.

(10.41) DEFINITION

A normal surface singularity $(X,0)$ is called *Gorenstein* if there is a 2-form ω on X' without any zeros (or, equivalently, the line bundle $\Omega^2_{X'}$ is trivial).

(10.42) EXERCISE

Let $X:f=0$ be an isolated surface singularity at the origin of $\mathbb{C}^3$. Show that the 2-form

$$\omega_X = \frac{dy\wedge dz}{\frac{\partial f}{\partial x}}$$

extends to a 2-form on $X'=X\setminus\{0\}$ without zeros. Hint: Use the relation

$$\frac{\partial f}{\partial x}dx+\frac{\partial f}{\partial y}dy+\frac{\partial f}{\partial z}dz=0$$

to extend ω_X over the set $\frac{\partial f}{\partial x}=0$.

Conclude that $(X,0)$ is a Gorenstein singularity.

(10.43) LEMMA

Let $\omega\in H^0(X',\Omega^2_{X'})$ *be a form without zeros. If* $p^*\omega$ *extends to a holomorphic form on* M, *then* $p_g(X,0)=0$.

PROOF

Any other 2-form η on X' can be written as $\eta=a\cdot\omega$ for some analytic function a on X'. Using the normality of $(X,0)$, it follows that a extends to an analytic function $\bar{a}$ on X, [KK], p. 315. If $\bar{\omega}$ denotes the extension of $p^*(\omega)$, then $p^*(\bar{a})\cdot\bar{\omega}$ is clearly an extension for $p^*(\eta)$ and hence, by our remark above, $p_g(X,0)=0$.

§5. SOME PROPERTIES OF SIMPLE SURFACE SINGULARITIES

An isolated surface singularity $X: f=0$ at the origin of $\mathbb{C}^3$ is called *simple* if f is a K-simple function germ (and hence f is K-equivalent to one of the normal forms A_k , D_k , E_6 , E_7 or E_8 in three variables from Theorem (8.26)).

Our aim in this section is to present some special interesting properties of these simple singularities, which in fact characterize them in the class of normal surface singularities. Our treatment is similar in spirit to [Du], but we have tried to keep it much more elementary.

(10.44) THEOREM

For an isolated surface singularity $X: f=0$ *at the origin of* $\mathbb{C}^3$ *the following statements are equivalent.*

(i) X *is a simple surface singularity.*

(ii) *The minimal resolution of* X *is very good and any component of its exceptional set is a rational curve with self--intersection* -2.

(iii) *The dual graph of the minimal resolution of* X *is one of the next weighted graphs (all weights associated to vertices being* -2) *and any component of its exceptional set is a rational curve.*

(iv) *The singularity* (X,0) *is rational and* mult (X,0)=2.

(v) *The singularity* (X,0) *is absolutely isolated and* mult (X,0)=2.

We point out that in the equivalence (i) $\Longleftrightarrow$ (iii) the type of the simple singularity X (given by the K-equivalence class of f listed in (8.26)) and the type of the corresponding dual resolution graph coincide. Moreover, these weighted graphs and their associated symbols A_k , D_k , E_6 , E_7 and E_8 are known as the classical *Dynkin diagrams of the simple complex Lie groups (or Lie algebras)* whose root systems have only roots of equal length, [H1], p. 58. A more direct connection between simple hypersurface singularities and simple Lie groups was conjectured by Grothendieck and proved by Brieskorn [B].

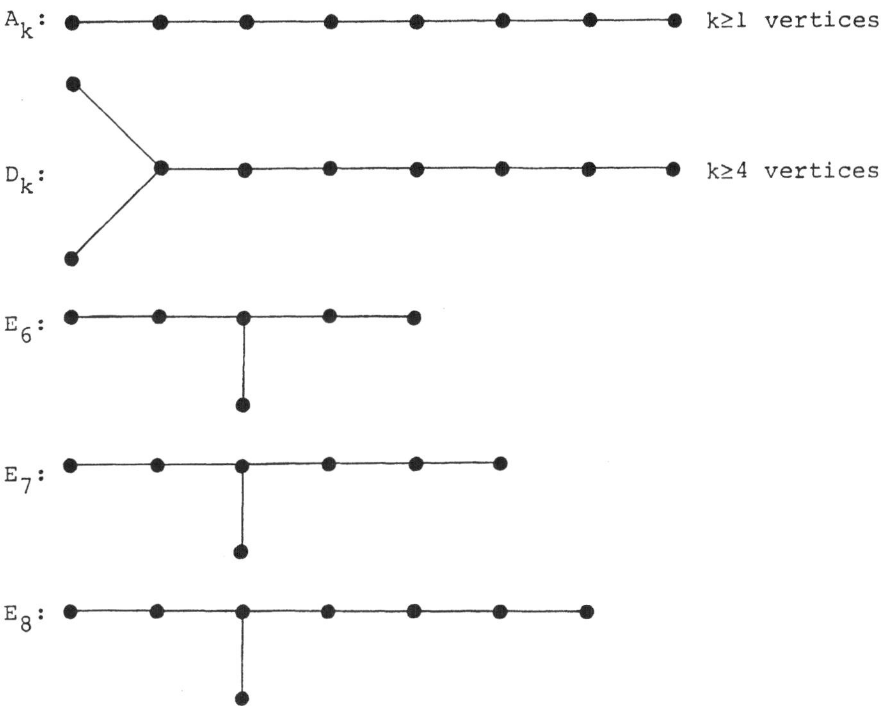

All this provide a sufficient motivation for choosing the symbols A-D-E to denote the simple hypersurface singularities.

We prove below only the following implications

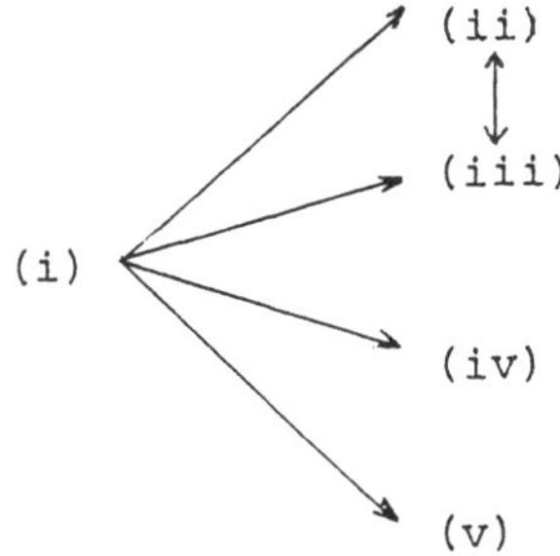

and send the reader interested in completing the proof to [Du] and [P].

PROOF

First note that (iii) $\Longrightarrow$ (ii) is obvious.

Next we prove (ii) $\Longrightarrow$ (iii). For any weighted graph G we consider a symmetric bilinear form $q^G : \mathbb{R}^s \times \mathbb{R}^s \to \mathbb{R}$, where s is the number of vertices in G, defined as follows.

We put $q^G(e_i, e_j) = -1$ if and only if the vertices v_i and v_j are connected by a segment in G and $q^G(e_i, e_i) = -w_i$, if the vertex v_i has weight w_i in G. Here $e_1, \dots, e_s$ denote the standard basis of $\mathbb{R}^s$.

If G is the dual graph of a resolution of a normal surface singularity, then G is connected and q^G is a positive definite form.

(10.45) LEMMA

The only connected graphs G with all vertices weighted by -2 and with positive definite associated forms q^G *are of the following type*

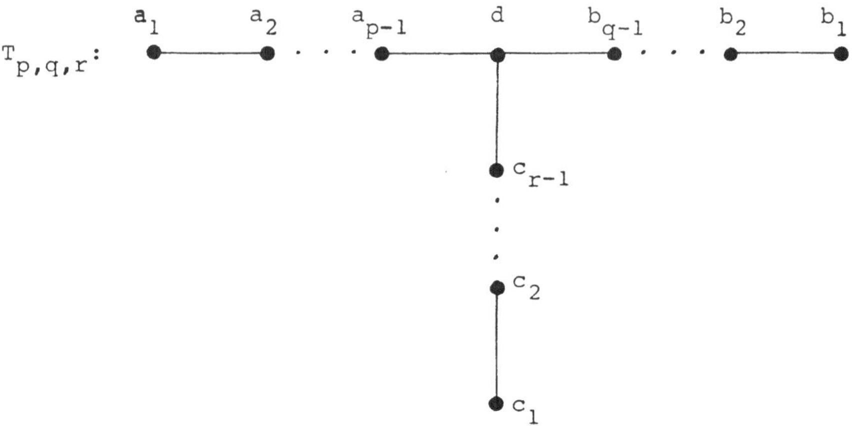

where the positive integers p, q, r *(the lengths plus one of the branches of this graph) satisfy*

$$\frac{1}{p} + \frac{1}{q} + \frac{1}{r} > 1.$$

Note that by symmetry one can assume that $1 \le p \le q \le r$ and then the only solutions of the above inequality are

(A) $p=1,\ r\geq q\geq 1$; then $T_{1,q,r}=A_{q+r-1}$

(D) $p=q=2,\ r\geq 2$; then $T_{2,2,r}=D_{r+2}$

(E) $p=2,\ q=3$ and $r=3,4,5$; then $T_{2,3,r}=E_{r+3}$.

This clearly is enough for proving the implication (ii)$\Longrightarrow$(iii) in Theorem (10.44).

PROOF OF THE LEMMA

STEP 1. If $H\subset G$ is a subgraph, then the associated form q^H is positive definite (being the restriction of q^G to some linear subspace).

STEP 2. G cannot contain cycles, i.e. subgraphs H of the form

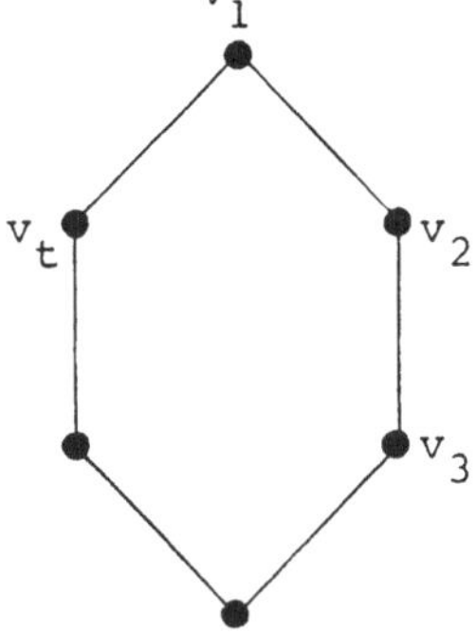

with $t\geq 3$ vertices. This follows from the equality

$$q^G(e_1+\ldots+e_t,\ e_1+\ldots+e_t)=0$$

STEP 3. A vertex $v\in G$ is called a *branch vertex* of order t if v is connected with $t\geq 3$ different vertices in G.

We show that a graph G satisfying the assumptions in (10.45) cannot have a branch vertex of order $t\geq 4$ or more than one branch vertex of order $t=3$.

Consider the subgraph $\widetilde{H}$ in G and note that the case $k=1$ corresponds to a branch vertex e_1 of order ≥ 4 and the case $k>1$ corresponds to two branch vertices e_1 and e_k of order 3. A direct computation shows that

$\tilde{H}$:

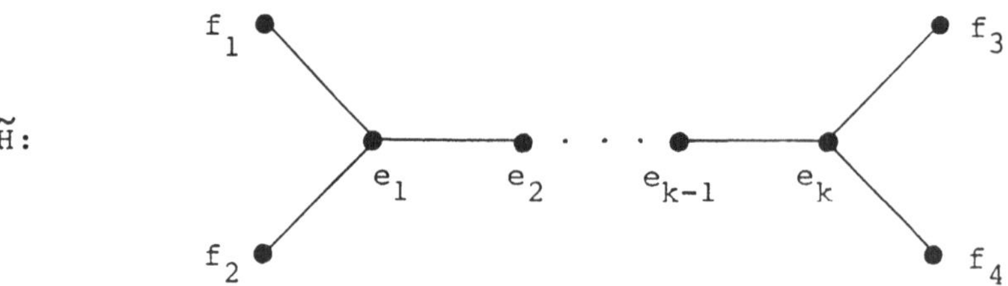

$$q^G(f_1+f_2+f_3+f_4+2e_1+\ldots+2e_k,\ f_1+f_2+f_3+f_4+2e_1+\ldots+2e_k)=0$$

which is a contradiction.

Here and in the sequel we identify a vertex in G with the corresponding basis vector in $\mathbb{R}^s$.

STEP 4. By the above steps, the graph G must be of type $T_{p,q,r}$ for some integers $p,q,r\geq 1$.

We use the notation for the vertices of G shown in the picture from the statement of Lemma (10.45). For simplicity of notation, we write

$$q^G(x,y)=\langle x,y\rangle,\quad q^G(x,x)=|x|^2 \quad \text{for} \quad x,y\in\mathbb{R}^s .$$

Consider the following three vectors in $\mathbb{R}^s$:

$$a=a_1+2a_2+\ldots+(p-1)a_{p-1}$$

$$b=b_1+2b_2+\ldots+(q-1)b_{q-1}$$

$$c=c_1+2c_2+\ldots+(r-1)c_{r-1}$$

A simple computation yields

$$|a|^2=p(p-1),\quad |b|^2=q(q-1),\quad |c|^2=r(r-1) .$$

Let V be the vector space spanned by a,b,c and let $\pi:\mathbb{R}^s\to V$ denote the orthogonal projection onto V, with respect to the scalar product $\langle\, ,\, \rangle$.

Since $d\notin V$, the well-known Pythagora's Theorem shows that

$|d|^2>|\pi(d)|^2$.

Next note that

$$\pi(d)=\langle d,\bar{a}\rangle\bar{a}+\langle d,\bar{b}\rangle\bar{b}+\langle d,\bar{c}\rangle\bar{c}$$

$$|\pi(d)|^2=\langle d,\bar{a}\rangle^2+\langle d,\bar{b}\rangle^2+\langle d,\bar{c}\rangle^2$$

where $\bar{a}=a\cdot|a|^{-1}$, $\bar{b}=b\cdot|b|^{-1}$, $\bar{c}=c\cdot|c|^{-1}$ are the associated unit vectors. Since

$$\langle d,\bar{a}\rangle^2=\frac{\langle d,a\rangle^2}{|a|^2}=\frac{(p-1)^2}{p(p-1)}=1-\frac{1}{p}$$

and similarly for $\langle d,\bar{b}\rangle^2$, $\langle d,\bar{c}\rangle^2$, the inequality from the statement follows using $|d|^2=2$.

STEP 5. By the remark following (10.45), it follows that G must be one of the graphs A_k $(k\geq 1)$, D_k $(k\geq 4)$, E_6, E_7 or E_8.

A direct computation in each of these cases (or the knowledge of some root system theory [H1]) shows that the corresponding forms q^G are indeed positive definite in all of these cases.

And this ends the proof of Lemma (10.45) and of the equivalence (ii)$\Longleftrightarrow$(iii) in (10.44).

Next we show that (i)$\Longrightarrow$(ii), (iii), (iv) and (v) in (10.44) by explicitly constructing the minimal resolution of the simple singularity (X,0) by means of successive blowing-ups of points.

We present only the case of A_k-singularities but with so many details that the interested reader will find no difficulty in treating similarly the other cases D_k, E_6, E_7 or E_8.

For the A_k-singularity we choose the defining function germ in normal form, namely we take

(10.46) $\quad X_k: x^{k+1}+y^2+z^2=0 \quad$ for $\quad k\geq 1$.

The construction of the minimal resolution of $(X_k,0)$ is by induction on k.

CASE k=1. This is a special case of Example (10.39), i.e. d=2 and hence (ii), (iii) and (v) follow directly.

To prove (iv) we use (10.43) and start with the 2-form on X_1 given in (10.42):

$$\omega_1 = \frac{dy \wedge dz}{x}$$

Recall that $Z=B_o(\mathbb{C}^3)$ is covered by 3 coordinate charts Z_i $i=1,2,3$ corresponding to $u_i \neq 0$.

In the open set Z_1 we have coordinates (x,u_2,u_3) such that the canonical projection $p:Z \to \mathbb{C}^3$ is given by $p(x,u_2,u_3)= =(x,u_2x,u_3x)$.

The exceptional divisor E is given by $x=0$, while the proper transform $\overline{X}_1$ has equation

$$(10.47) \qquad 1+u_2^2+u_3^2=0$$

If $q:\overline{X}_1 \to X_1$ is the resolution obtained by one blowing-up, then its exceptional set $q^{-1}(0)$ is precisely $E \cap \overline{X}_1 = \overline{E}$, a smooth conic in $\mathbb{P}^2$.

We have to show that $q^*(\omega_1)$, regarded as a 2-form on $\overline{X}_1 \setminus \overline{E}$ extends to a holomorphic form on $\overline{X}_1$. Since $Z_1 \cap \overline{E}$ is a Zariski open set in $\overline{E}$, it is enough to show that $q^*(\omega_1)|(Z_1 \cap \overline{X}_1)\setminus\overline{E}$ extends to $Z_1 \cap \overline{X}_1$. Now, on $(Z_1 \cap \overline{X}_1)\setminus\overline{E}$ we have the equality

$$q^*(\omega_1) = \frac{d(u_2x) \wedge d(u_3x)}{x} = u_3du_2 \wedge dx - u_2du_3 \wedge dx + xdu_2 \wedge du_3$$

And this clearly shows that $q^*(\omega_1)$ can be extended to $Z_1 \cap \overline{X}_1$. By (10.43) we get $p_g(X_1)=0$ and hence (iv) is proved.

Moreover, using (10.42) and the equality on $Z_1 \cap \overline{X}_1$

$$q^*(\omega_1) \wedge (u_2du_2+u_3du_3) = -dx \wedge du_2 \wedge du_3$$

of 3-forms on Z_1 , it follows that the extension of $q^*(\omega_1)$ to $\overline{X}_1$ has no zeros.

In other words, the line bundle $\Omega^2_{\overline{X}_1}$ is trivial.

CASE k=2. If $\overline{X}_2$ denotes the proper transform of the singularity X_2 under one blowing up, the equations for $\overline{X}_2 \cap Z_j$ $(j=1,2,3)$ are the following (using coordinates (y,u_1,u_3) on Z_2 and (z,u_1,u_2) on Z_3)

(10.48) $x+u_2^2+u_3^2=0;\ u_1^3y+1+u_3^2=0;\ u_1^3z+u_2^2+1=0$.

These equations imply that the proper transform $\bar{X}_2$ is smooth, i.e. $q:\bar{X}_2 \to X_2$, the blowing-up projection, is already a resolution.

The exceptional set $\bar{E}=E\cap\bar{X}_2=q^{-1}(0)$, regarded as a subset in $E=\mathbb{P}^2$ is given by the equation

$$u_2^2+u_3^2=0$$

Hence $\bar{E}$ consists of two lines $L_1:u_2+iu_3=0$ and $L_2: u_2-iu_3=0$ meeting transversally at the point $A=(1:0:0)\in E$. It follows that L_1, L_2 are the components of $\bar{E}$, they are rational curves and $(L_1, L_2)=1$. Before computing the self intersection numbers (L_j, L_j), we prove (iv).

We start with the 2-form on X_2 given by

$$\omega_2=\frac{dy\wedge dz}{x^2}$$

and a computation as above shows that on $(Z_1\cap\bar{X}_2)\setminus\bar{E}$ we have the equality

$$q^*(\omega_2)=\frac{u_3du_2\wedge dx-u_2du_3\wedge dx+xdu_2\wedge du_3}{x}$$

By (10.48) we get $dx+2u_2du_2+2u_3du_3=0$ and this implies

$$q^*(\omega_2)=3du_2\wedge du_3$$

which clearly gives the extension property.

Moreover the equality

$$q^*(\omega_2)\wedge(dx+2u_2du_2+2u_3du_3)=3du_2\wedge du_3\wedge dx$$

shows that the extension of $q^*(\omega_2)$ on $\bar{X}_2$ has no zeros and hence the line bundle $\Omega^2_{\bar{X}_2}$ is trivial.

Now we recall a basic result for a smooth curve C on a smooth surface M. Let K be the divisor on M corresponding to the zeros and poles of a rational (meromorphic) section of the line bundle Ω^2_M. Then one has the following

(10.49) ADJUNCTION FORMULA

$$g(C)=\frac{(C,C+K)}{2}+1$$

where g(C) *is the genus of the curve* C *and* (C,C+K) *denotes the intersection number of the divisors* C *and* C+K.

For details, we refer to [Ha], p. 361 or [BK], p. 624. In particular, if there is a section of Ω_M^2 which is holomorphic and has no zeros, then K=0 and it follows that

$$(10.50)\qquad (C,C)=2g(C)-2\ .$$

Coming back to our case, we get

$$(L_1,L_1)=(L_2,L_2)=-2$$

since $g(L_1)=g(L_2)=0$, L_j being lines isomorphic to $\mathbb{P}^1$. In this way all the assertions (ii), (iii), (iv) and (v) are proved in the case k=2.

Before passing to the general case, we make a remark useful for the induction.

Note that the lines

$$L_{\pm}: x=0 \quad y\pm iz=0$$

are contained in the affine surface X_k for all $k\geq 1$.

So it makes sense to speak about their proper transforms $\bar{L}_{\pm}$ in $\bar{X}_k$ for k=1,2.

For illustration, let us consider the case k=2. Working in the coordinate chart Z_2 , we get the following equations

$$\bar{L}_{\pm}\cap Z_2:\ u_1=0,\quad 1\pm iu_3=0$$

Since the exceptional divisor E is given by y=0, we obtain

$$\bar{L}_{\pm}\cap E=(0,0,\mp i)\text{ or, in longer notation }((0,0,0),(0:1:\mp i))\in Z.$$

More precisely, this means that we have the transverse intersections in $E=\mathbb{P}^2$

$\bar{L}_- \cap L_1 = (0:1:i)$, $\bar{L}_+ \cap L_2 = (0:1:-i)$.

We can describe this situation by the dual graph

$$\bar{L}_- \quad L_1 \quad L_2 \quad \bar{L}_+$$

(10.51) $\circ$———$\bullet$———$\bullet$———$\circ$

A similar discussion in the case k=1 leads to the next dual graph

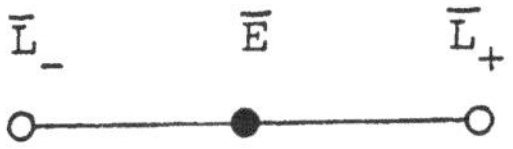

$$\bar{L}_- \quad \bar{E} \quad \bar{L}_+$$

$\circ$———$\bullet$———$\circ$

Now we statethe *induction hypothesis*.

Let $n \geq 3$ and for any $k<n$ we let $p_k : M_k \to X_k$ be the minimal resolution of the singularity $(X_k, 0)$. We assume that:

(a) M_k is obtained by a sequence of blowing-ups of points;

(b) The dual graph of the resolution M_k is of type A_k , all the components being rational curves;

(c) If ω_k denotes the 2-form

$$\omega_k = \frac{dy \wedge dz}{x^k}$$

defined on $X_k \setminus \{0\}$, then the pull-back $p_k^*(\omega_k)$ extends to a holomorphic 2-form on M_k without zeros;

(d) If $L_\pm$ denote the lines $x=0$, $y \pm iz = 0$ contained in X_k and if $\bar{L}_\pm$ denote their proper transforms under p_k , then the intersections among $\bar{L}_\pm$ and the components of the exceptional set $\bar{E}_k = p_k^{-1}(0)$ are described by the following dual graph:

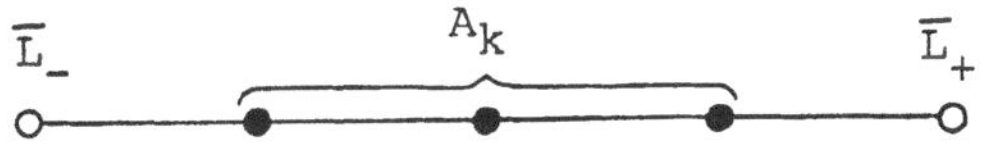

$$\bar{L}_- \qquad \overbrace{\quad\quad\quad}^{A_k} \qquad \bar{L}_+$$

$\circ$———$\bullet$———$\bullet$———$\bullet$———$\circ$

Here by proper transform under p_k we mean preciselv that $\bar{L}_\pm$ is the closure of $p_k^{-1}(L_\pm \setminus \{0\})$. Note also that in this hypothesis we deal with global algebraic varieties.

It is clear by the first two cases treated above that all the properties (a)-(d) are fulfilled for k=1,2.

Consider now the singularity X_n and let $\bar{X}_n$ denote the proper transform of X_n in Z (i.e. under one blowing-up of the origin).

The equation of $\bar{X}_n$ in the coordinate chart Z_1 is

(10.52) $(n+1)x^{n-1}+u_2^2+u_3^2=0$

and hence there is a singularity of type A_{n-2} at the point $A=(0,0,0)$ corresponding to $((0,0,0);(1:0:0))\in Z$.

Using the other coordinate charts, it is easy to see that this is the only singular point of $\overline{X}_n$. The exceptional set $\overline{E}=E\cap\overline{X}_n$ consists of two lines which in $E=\mathbb{P}^2$ are given by $u_2^2+u_3^2=0$ and in Z_1 are given by $x=0$, $u_2\pm iu_3=0$.

In other words $Z_1\cap\overline{E}$ corresponds exactly to the two lines $L_\pm$ considered in condition (d) above.

Let $q:\overline{X}_n\to X_n$ denote the canonical projection. Then a direct computation using (10.52) and

$$(n+1)(n-1)x^{n-2}dx+2u_2du_2+2u_3du_3=0$$

shows that, on the open set $(\overline{X}_n\cap Z_1)\setminus\overline{E}$ we have

$$q^*(\omega_n)=(1+\frac{2}{n-1})\frac{du_2\wedge du_3}{x^{n-2}} \tag{10.53}$$

It follows that $q^*(\omega_n)$ is well-defined on $\overline{X}_n\setminus\{A\}$ (use also the other coordinate charts!) and has no zeros. And around the singular point A, the form $q^*(\omega_n)$ coincides with the form ω_{n-2} up to an innocuous constant factor.

Let now $\overline{p}_{n-2}:\overline{M}_{n-2}\to\overline{X}_n$ be the minimal resolution of singularities for $\overline{X}_n$ constructed using the minimal resolution $p_{n-2}:M_{n-2}\to X_{n-2}$ of the singularity A_{n-2} applied to the point $A\in\overline{X}_n$. More precisely $\overline{M}_{n-2}$ is obtained by identifying the open subsets $M_{n-2}\setminus p_{n-2}^{-1}(0)$ and $\overline{X}_n\cap Z_1\setminus\{A\}$ in the disjoint union $M_{n-2}\cup\overline{X}_n$, via the isomorphism induced by p_{n-2} (this is a special case of a general construction called *glueing of two algebraic varieties*, [Ha], p. 75).

If we denote by p_n the composition $q\circ\overline{p}_{n-2}$, it follows easily that $p_n:M_n=\overline{M}_{n-2}\to X_n$ is a minimal resolution of singularities for X_n satisfying (a)-(b)-(c).

For instance, the end points in the string

A_n ●———●———●———●———●

correspond precisely to the proper transforms of the (projective) lines $L_\pm \subset X_n$ under p_{n-2} by the condition (d). These proper transforms have self-intersection numbers -2 by the formula (10.50) which can be applied by condition (c) and since these curves are rational (being isomorphic to $L_\pm$ via p_{n-2}).

Next the condition (d) for p_n is also fulfilled. This can be done by a direct computation in the coordinate charts, similar to that which has led to the dual graph (10.51).

And this ends our (partial, but we hope quite instructive!) proof of Theorem (10.44).

(10.54) EXERCISE

Compute the minimal resolutions (at least) for the simple surface singularities of type D_4 and E_6. Use this to show that these singularities are rational as in the above proof.

(10.55) REMARK

Let X and Y be two simple surface singularities.

Then a case-by-case analysis of the diagram of specializations (8.24) shows the following interesting property:

The singularity X *specializes to* Y *if and only if* D_X *is a subgraph in* D_Y, *where* D_X, D_Y *are the corresponding dual resolution graphs (Dynkin diagrams).*

In notation: $X \to Y \iff D_X \subset D_Y$ and this is one reason for prefering the notation $X \to Y$ to the notation $X \leftarrow Y$.

A deeper explanation of this remark relies on Milnor lattices and can be found in full generality in [Ln], Chapter 7.

Finally we remark that different other constructions for the minimal resolutions of simple surface singularities can be found for instance in [La], Chap. II and [Lm], Chap. IV. This last reference contains also a detailed description of simple surface singularities as quotient singularities of $\mathbb{C}^2$ under finite groups.

CHAPTER 11

DUAL MAPPINGS AND CONTACT TANGENCY CLASSES FOR PROJECTIVE HYPERSURFACES

§1. THE DUAL MAPPING AND THE DUAL VARIETY

Let $V:f=0$ be a smooth hypersurface in the complex projective space $\mathbb{P}^n$, defined by a homogeneous polynomial $f\in\mathbb{C}[x_0,\dots,x_n]$ of degree $d\geq 2$.

The *projective tangent space* of V at a point $a\in V$ is the hyperplane in $\mathbb{P}^n$ given by

$$(11.1)\qquad T_aV:\frac{\partial f}{\partial x_0}(a)x_0+\dots+\frac{\partial f}{\partial x_n}(a)x_n=0$$

The set of all hyperplanes in $\mathbb{P}^n$ forms the *dual projective space* $\hat{\mathbb{P}}^n$ and the map

$$(11.2)\qquad \varphi:V\to\hat{\mathbb{P}}^n\ ,\quad \varphi(a)=T_aV$$

is called the *dual mapping* of the hypersurface V. Moreover, its image $\hat{V}=\mathrm{im}\,(\varphi)$ is a (singular) hypersurface in $\hat{\mathbb{P}}^n$ (see (11.5) below) and it is called the *dual variety* (or *dual hypersurface*) of V. For the case $n=2$, one can see [BK], p. 251 for a more detailed discussion.

The purpose of this Chapter is to investigate the singularities of the dual mapping φ and of the dual variety $\hat{V}$ and to relate them to the hypersurface singularities

$$(11.3)\qquad (V\cap T_aV,\ a)\subset(T_aV,\ a)\simeq(\mathbb{C}^{n-1},\ 0)\quad \text{for}\quad a\in V.$$

We denote (the isomorphism class of) this hypersurface singularity by $K_a(V)$ and call it the *contact tangency class* of the hypersurface V at the point a. Clearly, $K_a(V)$ describes the *order of contact* of V with its tangent hyperplane T_aV at the point a.

Let $H\in\hat{\mathbb{P}}^n$ be a hyperplane. Then $a\in V\cap H$ is a singular point for this intersection (called a *hyperplane section*) if and only if $H=T_aV$. It follows that the dual mapping φ is finite ($\#\varphi^{-1}(H)<\infty$ for any $H\in\hat{\mathbb{P}}^n$) if and only if all the singularities $K_a(V)$ are isolated hypersurface singularities.

Our first result shows that this is indeed the case.

(11.4) PROPOSITION

Let V *be a (not necessarily) smooth hypersurface in* $\mathbb{P}^n$ *and* $V_o=V\cap H$ *one of its hyperplane sections. Then*

$$\dim S(V)-1\leq\dim S(V_o)\leq\dim S(V)+1$$

Here S(X) denotes the singular set of an algebraic variety X and $\dim\emptyset=-1$ by convention.

PROOF

The inclusion $H\cap S(V)\subset S(V_o)$ proves the first inequality.

To prove the second one, we choose coordinates on $\mathbb{P}^n$ such that

$$H:x_o=0\ ,\quad V:f=0$$

Then $V_o:f(0,x_1,\dots,x_n)=0,\ x_o=0$. And $S(V_o)$ is defined by the system of equations in H:

$$\frac{\partial f}{\partial x_i}(0,x_1,\dots,x_n)=0\quad\text{for}\quad i=1,\dots,n\ .$$

Consider also the hypersurface W in H given by

$$W:\frac{\partial f}{\partial x_o}(0,x_1,\dots,x_n)=0$$

Then clearly $W\cap S(V_o)\subset S(V)$ and hence

$$\dim S(V)\geq\dim(W\cap S(V_o))\geq\dim S(V_o)-1$$

(11.5) COROLLARY

If V *is a smooth hypersurface in* $\mathbb{P}^n$*, then any hyperplane* $H\subset P^n$ *is tangent to* V *at a finite number of points (possible zero).*

Equivalently, the dual mapping is finite and all the con-

tact tangency classes $K_a(V)$ *are isolated hypersurface singularities.*

In particular, it follows that $\dim \hat{V} = \dim V = n-1$ *and hence* $\hat{V}$ *is a hypersurface in* $\hat{\mathbb{P}}^n$.

Next we prove a converse to this Corollary.

(11.6) PROPOSITION

Let H *be a hyperplane in* $\mathbb{P}^n$ *and* V_o *be a hypersurface in* H *having at most isolated singularities. Then there exists a smooth hypersurface* $V \subset \mathbb{P}^n$ *such that* $V_o = V \cap H$.

PROOF

We can choose the linear coordinate system on $\mathbb{P}^n$ such that $H: x_o = 0$.

Let $V_o: g(\bar{x}) = 0$ be the equation of V_o in H, where g is a homogeneous polynomial in $\bar{x} = (x_1, \dots, x_n)$ of degree $d \geq 2$. (The case $d=1$ is never considered in this Chapter since all the statements made here are either trivially true or trivially false in this case!)

We look for a polynomial $f = x_o f_o(x) + g(\bar{x})$ such that $V: f=0$ is smooth. We take

$$f_o(x) = \frac{1}{d-1} \sum_{i=0,n} a_i x_i^{d-1}$$

and prove that there exists $a = (a_o, \dots, a_n) \in \mathbb{C}^{n+1}$ such that the corresponding hypersurface V is smooth.

Note that $x \in S(V)$ if and only if x is a solution of the system of equations

$$S: \begin{cases} f_o(x) + x_o \dfrac{\partial f_o}{\partial x_o}(x) = 0 \\ x_o \dfrac{\partial f_o}{\partial x_i}(x) + \dfrac{\partial g}{\partial x_i}(\bar{x}) = 0 \qquad \text{for } i=1,\dots,n . \end{cases}$$

First, if $x = (x_o, \bar{x}) \in S(V)$ and $x_o = 0$, it follows that $\bar{x} \in S(V_o)$. Since $S(V_o)$ is a finite set, there are hyperplanes $H_1, \dots, H_p$ in $\mathbb{C}^{n+1}$ such that for $a \in \mathbb{C}^n \setminus \bigcup_{i=1,p} H_i$ we have $f_o(0, \bar{x}) \neq 0$

for all $\bar{x}\in S(V_o)$.

Hence, for this choice of the coefficients a, the system S has no solutions with $x_o=0$.

Assume now that $x\in S(V)$ and $x_o\neq 0$. Then the system S is equivalent to the system

$$S': a_i x_i^{d-2} = -\frac{\partial}{\partial x_i}\left(\frac{g(x)}{x_o}\right) \quad \text{for} \quad i=0,1,\dots,n.$$

For any subset $I=\{i_1<\dots<i_q\}\subset\{0,1,\dots,n\}$ with $i_1=0$ we define a map

$$f_I:(\mathbb{C}^*)^q \to \mathbb{C}^q,\ f_I^k(t_1,\dots,t_q) = -\frac{\partial}{\partial x_{i_k}}\left(\frac{g}{x_o}\right)(\bar{t})\cdot t_k^{2-d}$$

Here $k=1,\dots,q$ and the point $\bar{t}$ in $\mathbb{C}^{n+1}$ is given by $\bar{t}_i=t_e$ for $i=i_e$ and $\bar{t}_j=0$ if $i\neq i_e$ for all $e=1,\dots,q$.

Note that f_I^k is a quotient of homogeneous polynomials of degree d when k=0 and of degree d-1 when k>0.

Hence we get a commutative diagram

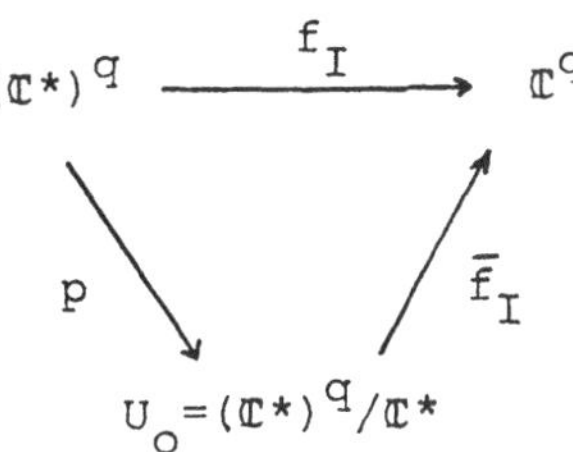

where U_o is an open subset in $\mathbb{P}^{q-1}=\mathbb{P}(\mathbb{C}^q)$ and $\bar{f}_I$ is a morphism of algebraic varieties.

It follows that $K_I=\operatorname{im} f_I$ is a semialgebraic set of dimension $\leq q-1$.

Let $p_I:\mathbb{C}^{n+1}\to\mathbb{C}^q$ be the projection

$$p_I(x)=(x_{i_1},\dots,x_{i_q})$$

It is now easy to see that for any coefficients $a\in\mathbb{C}^{n+1}\setminus\left(\bigcup_i H_i\cup\bigcup_I L_I\right)$, with $L_I=p_I^{-1}(K_I)$ the corresponding hypersurface V:f=0 is smooth.

(11.7) REMARK

The results (11.4) and (11.6) hold in fact for the larger class of smooth complete intersections, see for instance [Is].

It follows from (11.4) and (11.6) that *finding the possible contact tangency classes* $K_x(V)$ *for* x *a point on a smooth hypersurface* V *in* $\mathbb{P}^n$ *of degree* d *is equivalent to finding the possible singularities of a hypersurface* V_o *in* $\mathbb{P}^{n-1}$ *of degree* d, *having only isolated singularities.*

(11.8) EXERCISE

(i) For smooth hyperquadrics ($d=2$, $n\geq 2$) the only possible class $K_x(V)$ is A_1.

(ii) For smooth plane curves of degree d, the possible classes $K_x(V)$ are A_i with $1\leq i<d$.

(iii) For smooth cubic surfaces ($n=d=3$), one has $K_x(V)\in \in\{A_1,A_2,A_3,D_4\}$. Hint: use the classification of singular cubic curves (5.16).

When we have classified the smooth cubic curves in Chapter 5 an important role was played by their inflexion points. It is intuitively clear that there is some relation between the inflexion points, the dual mapping and the contact tangency classes of a given smooth hypersurface V in $\mathbb{P}^n$. This is precisely what we are going to explain now.

Let f be a homogeneous polynomial of degree $d\geq 2$ in $x_o,\ldots,x_n$ and let F be the associated *gradient map*:

$$F:\mathbb{C}^{n+1}\to\mathbb{C}^{n+1}\quad,\quad F(x)=\left(\frac{\partial f}{\partial x_o}(x),\ldots,\frac{\partial f}{\partial x_n}(x)\right)$$

We denote by $h(f)(x)$ the Jacobian matrix of this map F at the point x, namely

$$h(f)(x)=\left(\frac{\partial^2 f}{\partial x_i\partial x_j}(x)\right)_{0\leq i,j\leq n}$$

and put $H(f)(x)=\det(h(f)(x))$ for the hessian of the polynomial f at the point x as in (5.12).

Note that $x \in \mathbb{C}^{n+1}$ is a singular point for F if and only if $H(f)(x)=0$.

Assume from now on that the hypersurface $V: f=0$ defined by our polynomial f in the projective space $\mathbb{P}^n$ is smooth.

Then the gradient map F has a restriction

$$F: \mathbb{C}^{n+1}\setminus\{0\} \to \mathbb{C}^{n+1}\setminus\{0\}$$

which induces an algebraic morphism $\bar{F}: \mathbb{P}^n \to \mathbb{P}^n$ such that the following diagram commutes

$$(11.9)\qquad \begin{array}{ccc} \mathbb{C}^{n+1}\setminus\{0\} & \xrightarrow{F} & \mathbb{C}^{n+1}\setminus\{0\} \\ p\downarrow & & \downarrow p \\ \mathbb{P}^n & \xrightarrow{\bar{F}} & \mathbb{P}^n \end{array}$$

where p denotes the natural projection.

A moment thought shows that the differential

$$dp(x): T_x(\mathbb{C}^{n+1}\setminus\{0\}) \to T_{\bar{x}}\mathbb{P}^n$$

with $\bar{x}=p(x)$, induces an isomorphism

$$(11.10)\qquad \ker dF(x) \simeq \ker d\bar{F}(\bar{x}) \quad \text{for any} \quad x \in \mathbb{C}^{n+1}\setminus\{0\}.$$

Hint: use the fact that ker dp(x) is equal to the line $\mathbb{C}\cdot x$ and consider the restriction $F|\mathbb{C}\cdot x$.

Note that the dual mapping φ of V equals the composition $\bar{F}\circ j$, where $j: V \to \mathbb{P}^n$ is the inclusion. This observation gives us the first indication about the singularities of the dual mapping.

(11.11) LEMMA

$\ker d\varphi(\bar{x}) = \ker d\bar{F}(\bar{x})$ *for any point* $x \in V$.

PROOF

It is enough to show the inclusion

$$\ker d\overline{F}(\overline{x}) \subset T_{\overline{x}}V$$

where we let here $T_{\overline{x}}V$ denote the linear tangent space and not the associated projective hyperplane.

Let $\overline{v} \in \ker d\overline{F}(\overline{x})$ and $v \in \ker dF(x)$ be two corresponding vectors under the isomorphism (11.10). It follows that

$$dF(x)(v) = h(f)(x) \cdot v = 0$$

or, in other words:

$$\sum_{i=0,n} \frac{\partial^2 f}{\partial x_i \partial x_j}(x) v_i = 0 \qquad \text{for} \quad j=0,\dots,n. \tag{11.12}$$

Using the Euler relation (7.6) we get

$$\sum_{j=0,n} x_j \frac{\partial^2 f}{\partial x_i \partial x_j}(x) = (d-1) \cdot \frac{\partial f}{\partial x_i}(x)$$

It follows that (11.12) implies the equality

$$\sum_{i=0,n} \frac{\partial f}{\partial x_i}(x) \cdot v_i = 0 \tag{11.13}$$

If we denote by $\tilde{V}$ the *affine* hypersurface defined by the equation $f=0$ in $\mathbb{C}^{n+1}$ (the cone CV over V!), then the equality (11.13) tells us that $v \in T_x\tilde{V}$.

The differential $dp(x)$ induces a surjection $T_x\tilde{V} \to T_{\overline{x}}V$ and hence $\overline{v} = dp(x)(v) \in T_{\overline{x}}V$ as we claimed.

Let $H(V): H(f)=0$ be the *hessian hypersurface* associated to the smooth hypersurface V. The *set of the inflexion points* of V is by definition the intersection

$$\text{Inf}\,(V) = V \cap H(V)\ . \tag{11.14}$$

(11.15) COROLLARY

With the above notations, one has

$$\dim(\ker d\varphi(\overline{x})) = \text{corank}\, h(f)(x) = n+1-\text{rank}\, h(f)(x)\ .$$

In particular, the dual mapping φ *is an immersion at* $\overline{x}$ *if and only if* $\overline{x} \notin \text{Inf}\,(V)$.

PROOF

Use Lemma (11.11) and the isomorphism (11.10).

Two (nontrivial?) consequences of this result are the following.

(i) Inf $(V)\neq V$; indeed, a finite mapping between smooth algebraic varieties is obviously an immersion on some open and dense Zariski subset of its source, see [Mu], p. 42.

(ii) rank $h(f)(x)\geq 2$; indeed one clearly has $\dim(\ker d\varphi(x))\leq n-1$.

(11.16) EXERCISE

Show that for V a smooth (hyper)quadric (d=2) the dual mapping φ is an isomorphism. In particular Inf $(V)=\emptyset$ and the dual variety $\hat{V}$ is also a smooth (hyper)quadric.

In the sequel, we will avoid this completely special case by assuming $d=\deg(f)=\deg V\geq 3$.

We consider now a partition of the set of singular points for the dual mapping φ (equal to Inf (V)), namely we put

$$S_r(V)=\{x\in V;\ \dim(\ker d\varphi(x))=r\}$$

This is nothing else but the *first order Thom-Boardman partition associated to* φ, see (9.7) and [GG], p. 143. Let $H=H^2(n+1,1;\mathbb{C})$ be the vector space of all quadratic forms $\mathbb{C}^{n+1}\to\mathbb{C}$, $\mathbb{P}(H)$ the associated projective space and let

$$\psi:V\to\mathbb{P}(H),\qquad \psi(x)=h(f)(x)$$

be the *hessian mapping of* V, where we identify a symmetric $(n+1)\times(n+1)$-matrix with a quadratic form in H.

Then, as in Chapter 5, §3, there is a partition $\bigcup_{r=0,n} W_r=\mathbb{P}(H)$, where W_r is the smooth orbit corresponding (this time!) to the quadratic forms of *corank* $r, r=0,\dots,n$. Then by (11.15) we obtain the equality

$$(11.17)\qquad S_r(V)=\psi^{-1}(W_r)\quad \text{for any } r=0,\dots,n-1\ .$$

This relation can be used to obtain some interesting information on the semialgebraic sets $S_r(V)$. For instance, one has the next two results (for proofs and further details we refer to [CD]).

(11.18) PROPOSITION

For any smooth hypersurface V *the hessian mapping* ψ *is finite. In particular, the union* $\bigcup_{j\geq r} S_j(V)$ *is a closed nonempty variety of codimension* $\leq r(r+1)/2$ *in* V, *for any positive integer* r *such that*

(*) $\qquad r(r+1)/2\leq n-1.$

(11.19) PROPOSITION

For a generic smooth hypersurface V *the hessian mapping* ψ *is transverse to all the orbits* W_j. *In particular, in this situation* $S_r(V)$ *is a smooth nonempty subvariety of codimension* $r(r+1)/2$ *in* V, *for any positive integer* r *satisfying the condition* (*) *above.*

(11.20) COROLLARY

For a generic plane curve $V\subset \mathbb{P}^2$ *of degree* d, *the inflexion set* Inf (V) *consists of exactly* 3d(d-2) *points.*

PROOF

By Bezout's Theorem [BK], p. 232 it follows that the intersection number (V, H(V)) equals $\deg V\cdot\deg H(V)=d\cdot 3(d-2)$. And by (11.19) it follows that each point in the intersection $V\cap H(V)$ has multiplicity one, i.e.

$$(V, H(V))_x=1 \quad \text{for any} \quad x\in V\cap H(V)$$

if V is a generic curve. See also (11.46) below.

In the end of this section we would like to say a few words on the dual hypersurface $\hat{V}$ of our smooth hypersurface V in $\mathbb{P}^n$.

(11.21) PROPOSITION

A point $y \in \hat{V}$ *is nonsingular if and only if* $\varphi^{-1}(y)=\{x\}$ *and* $x \notin \mathrm{Inf}(V)$.

PROOF

From Sard's Theorem [Mu], p. 42 it follows that there is a dense Zariski open subset $W \subset \hat{V} \setminus S(\hat{V})$ such that φ is an immersion at any point $x \in U = \varphi^{-1}(W)$.

We have the following diagram

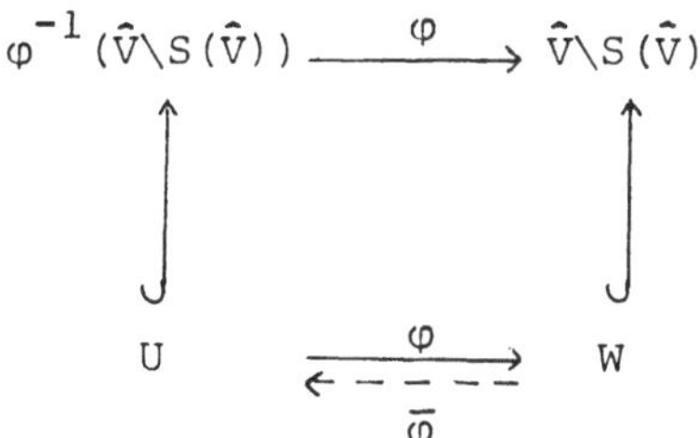

where $\bar{\varphi}(y)=(\frac{\partial g}{\partial y_o}(y),\ldots,\frac{\partial g}{\partial y_n}(y))$, with $g=0$ being a reduced equation for the hypersurface $\hat{V}$ in $\hat{\mathbb{P}}^n$. In other words, $\bar{\varphi}$ is a partial dual mapping for $\hat{V}$. Then $\bar{\varphi}$ is defined in fact on $\hat{V} \setminus S(\hat{V})$, the smooth part of $\hat{V}$, and a simple computation shows that $\bar{\varphi} \circ \varphi =$ $=$identity on U.

It follows that we also have $\bar{\varphi} \circ \varphi = 1$ on $\varphi^{-1}(\hat{V} \setminus S(\hat{V}))$. This implies that for $y \in \hat{V} \setminus S(\hat{V})$ we have $\varphi^{-1}(y)=\{x\}$ and φ is an immersion at the point x, i.e. $x \notin \mathrm{Inf}(V)$.

The converse implication is obvious.

(11.22) COROLLARY

For any smooth hypersurface V *one has the next.*

(i) *There is a Zariski open and dense subset* $U \subset V$ *such that the hyperplane section* $V \cap T_x V$ *for* $x \in U$ *has a unique singular point (at* x*) and this is of type* A_1.

(ii) $\operatorname{codim}_{\hat{V}} S(\hat{V})=1$ *and hence* $\hat{V}$ *is not a normal variety, as soon as* $\deg(V) \geq 3$.

PROOF

For (i) just take $U=\varphi^{-1}(\hat{V}\setminus S(\hat{V}))$. For (ii), note that $S(\hat{V})\supset\varphi(\text{Inf }(V))$ and $\dim \text{Inf }(V)=\dim V-1$. Then use (11.5) and (10.3).

(11.23) REMARK

In fact, the dual mapping $\varphi:V\to\hat{V}$ is precisely the normalization of $\hat{V}$. In addition, $\deg \hat{V}=d(d-1)^{n-1}$, see for instance [K1].

A useful generalization of Proposition (11.21) is the following. Let $y\in\hat{V}$ and let $x_1,\dots,x_p$ be the points in $\varphi^{-1}(y)$. Then we have p isolated hypersurface singularities $K_{x_i}(V)$, given by these contact tangency classes as defined in (11.3).

Then the multiplicity $\text{mult}_y(\hat{V})$ of the dual variety at the point y (defined to be the multiplicity of the hypersurface germ $(\hat{V},y)$ as in (7.48)) can be expressed in terms of the Milnor numbers of the singularities $K_{x_i}(V)$ in the following way.

(11.24) PROPOSITION

With the above notations

$$\text{mult}_y(\hat{V})=\sum_{i=1,p}\mu(K_{x_i}(V))$$

PROOF

Since this result is proved in more general conditions in [D4], we just give below a sketch of the proof (enough however for the reader who wants to fill in the details himself!)

Let V_i be a small open neighbourhood of x_i in V. Then $(\varphi(V_i),y)$ for $i=1,\dots,p$ are just the irreducible components of the germ $(\hat{V},y)$ and hence

$$\text{mult}_y(\hat{V})=\text{mult }(\hat{V},y)=\sum_{i=1,p}\text{mult }(\varphi(V_i),y)$$

Let $\hat{V}_i=\varphi(V_i)$ and $\varphi_i=\varphi|V_i$. For a nonsingular point $z\in\hat{V}_i$, it follows from (11.21) that $x=\varphi^{-1}(z)$ corresponds to the tangent hyperplane $T_z\hat{V}_i$ via the identification $\hat{\mathbb{P}}^n=\mathbb{P}^n$.

It follows that φ_i is bijective outside $S(\hat{V}_i)$ and that the limit

$$T_i=\lim T_z\hat{V}_i \text{ , for } z \to y,\ z\in\hat{V}_i\setminus S(\hat{V}_i)$$

exists and is equal to x_i via the above identification.

If $q_i:(\hat{V}_i,y) \to (T_i,y)$ denotes the orthogonal projection on the hyperplane germ (T_i,y), then the *geometric description of multiplicity* (see [Wt], p. 234) implies

$$\text{mult } (\hat{V}_i,y)=\deg q_i$$

where $\deg q_i$ denotes the number of sheets of the ramified covering q_i.

The composition $\bar{q}_i=q_i\circ\varphi_i$ is again a ramified covering (this time between smooth germs!) and obviously $\deg \bar{q}_i=\deg q_i$.

Using now the topological description of the Milnor number [Lm], p. 196 and a simple local computation of the covering $\bar{q}_i$ in terms of the singularity $K_{x_i}(V)$ similar to the proof of (11.30) below the result follows.

For a generic hypersurface V, the singularities of the dual hypersurface $\hat{V}$ can be described sometimes quite precisely in terms of *discriminant singularities*, see [Bc].

§2. CONTACT TANGENCY CLASSES

The definition (11.3) of the contact tangency classes $K_a(V)$ which we have given above is part of a more general situation. We describe briefly this situation in the beginning of this section for two reasons: to give the reader a complete picture and to explain the *geometric meaning of the contact tangency classes*.

Consider the n-dimensional smooth germ $(K^n,0)$, with $K=\mathbb{R}$ or $\mathbb{C}$ as usual. Suppose we are given two pairs $(X_1,0)$, $(Y_1,0)$ and $(X_2,0)$, $(Y_2,0)$ of smooth submanifold germs at the origin of K^n.

(11.25) DEFINITION

We say that the pair $(X_1,0)$, $(Y_1,0)$ has the same *contact* at the origin as the pair $(X_2,0)$, $(Y_2,0)$ if there is a diffeomorphism germ $h:(K^n,0) \to (K^n,0)$ such that $h(X_1,0)=(X_2,0)$ and $h(Y_1,0)=(Y_2,0)$.

In the complex case, we mean here that the submanifols are complex analytic submanifolds and that h is a complex analytic isomorphism germ.

To each of the pairs $(X_i,0)$, $(Y_i,0)$ above we can associate a map germ f_i in the following (natural but not unique) way.

Let $h_i:(K^{m_i},0) \to (X_i,0)$ be a parametrization (chart) for the manifold X_i around the origin, with $m_i=\dim X_i$. And let $g_i:(K^n,0) \to (K^{p_i},0)$ be a submersion map germ such that $(g_i^{-1}(0),0)=(Y_i,0)$ where $p_i=\operatorname{codim} Y_i$.

Note that such germs h_i and g_i exist indeed in all cases. Then we put $f_i=g_i \circ h_i$.

If the pairs $(X_1,0)$, $(Y_1,0)$ and $(X_2,0)$, $(Y_2,0)$ have the same contact, then one has in particular $m_1=m_2$ and $p_1=p_2$ and hence the map germs f_1 and f_2 can be compared.

More precisely, one has the next basic result.

(11.26) PROPOSITION

The pairs $(X_1,0)$, $(Y_1,0)$ *and* $(X_2,0)$, $(Y_2,0)$ *have the same contact at the origin if and only if* $m_1=m_2$, $p_1=p_2$ *and the map germs* f_1 *and* f_2 *are K-equivalent.*

The proof of this result in the real smooth case can be found in [Mo]. However, the reader can check easily that all the details of the proof work equally well in the complex case. He should also check that our definition of the contact tangency class $K_a(V)$ is a special case of the construction of the map germ f_i above.

Now we come back to the main theme of this Chapter. Let x be a point on the smooth hypersurface V in $\mathbb{P}^n$. We show that the contact equivalence class of the map germ

$$\varphi_x:(V,x) \to (\hat{\mathbb{P}}^n,\varphi(x))$$

induced by the dual mapping φ at the point x is completely determined by the contact tangency class $K_x(V)$.

Let $g \in K_x(V)$ be a representative for this K-equivalence class. Then $g:(\mathbb{C}^{n-1},0) \to (\mathbb{C},0)$ has an isolated singularity at the origin.

For any such function germ g we consider the map germ

$$\text{(11.27)} \quad t(g):(\mathbb{C}^{n-1},0) \to (\mathbb{C}^n,0), \quad t(g)=(g,\frac{\partial g}{\partial u_1},\dots,\frac{\partial g}{\partial u_{n-1}})$$

where $u_1,\dots,u_{n-1}$ are the standard coordinates on $\mathbb{C}^{n-1}$.

There is a close relation between the map germ t(g) and the Tjurina algebra T(g) associated to g as in (6.49). And the Mather-Yau Theorem [MY] discussed there can be stated as follows.

(11.28) THEOREM

Two finitely determined function germs $g,g' \in m_{n-2}^2$ *are* K*-equivalent if and only if the corresponding map germs* t(g), t(g') *are* K*-equivalent.*

Many contact invariants of the germs $g \in m_{n-1}^2$ and $t(g) \in$ $\in E^o_{n-1,n}$ are closely related. For instance we have the next easy result.

(11.29) EXERCISE

(i) corank (g)=dim (ker dt(g)(0))

(ii) If the Boardman symbol of the function germ g is $(i_1,i_2,i_3,\dots)$, then $i_1=n-1$ and the Boardman symbol of the map germ t(g) is $(i_2,i_3,\dots)$.

It follows from the easy part of (11.28) that one can associate to the K-equivalence class $K_x(V)$ a well-defined K-equivalence class of map germs in $E^o_{n-1,n}$, denoted by $t(K_x(V))$. With these notations, the result we are looking for is the following.

(11.30) PROPOSITION

The contact class of the dual mapping germ

$\varphi_x:(V,x) \to (\hat{\mathbb{P}}^n,\varphi(x))$

is equal to $t(K_x(V))$.

PROOF

By a suitable choice of the linear coordinates on $\mathbb{P}^n$ we can assume that $x=(1:0:\ldots:0)$ and $T_xV:x_n=0$. We can write V locally as the graph of a function, i.e. there is a function germ $g\in m_{n-1}^2$ such that

(11.31) $\quad f(1,u_1,\ldots,u_{n-1},g(u))=0$

for $u=(u_1,\ldots,u_{n-1})$ in a neighbourhood of the origin in $\mathbb{C}^{n-1}$. And we obviously have $g\in K_x(V)$.

We consider the isomorphism germs

$h:(\mathbb{C}^{n-1},0) \to (V,x)$, $h(u)=(1:u:g(u))$

$k:(\hat{\mathbb{P}}^n,\varphi(x)) \to (\mathbb{C}^n,0)$, $k(y_o:\ldots:y_n)=(\frac{y_o}{y_n},\ldots,\frac{y_{n-1}}{y_n})$

If $\tilde{\varphi}=k\circ\varphi_x\circ h$, then a simple computation using (11.31) shows that

$$\tilde{\varphi}(u)=-(g(u)-\sum_{i=1,n-1} u_i\frac{\partial g}{\partial u_i}(u),\ \frac{\partial g}{\partial u_1}(u),\ldots,\frac{\partial g}{\partial u_{n-1}}(u))$$

This obviously shows that the map germ φ_x (which in the local coordinates induced by h and k is expressed by the germ $\tilde{\varphi}$) belongs to the contact class $t(K_x(V))$. Indeed, the ideal generated by the components of $\tilde{\varphi}$ is equal to the ideal generated by the components of $t(g)$ and hence $\tilde{\varphi}$ is K-equivalent to $t(g)$ by Proposition (3.16).

(11.32) COROLLARY

corank $K_x(V)=\dim\ (\ker d\varphi(x))$.

We are going next to study in more detail the simplest

contact tangency classes, namely those of type A_1 and A_2, and relate them with the properties of the dual mapping.

In the first case, our results so far show that

(11.33) $K_x(V)=A_1 \Longleftrightarrow x\in V\setminus \mathrm{Inf}\,(V) \Longleftrightarrow \varphi_x$ is an immersion

The second case is more subtle and is described by the next result.

(11.34) PROPOSITION

A point $x\in V$ *satisfies* $K_x(V)=A_2$ *if and only if* $\mathrm{Inf}(V)$ *is a smooth complete intersection at* x *and* $\varphi_x|\mathrm{Inf}\,(V)$ *is an immersion.*

Before giving the proof, we note that the condition on the inflexion set Inf (V) means that the hessian hypersurface H(V) is nonsingular at x and in addition $T_xV\neq T_xH(V)$.

PROOF

Recall the notations used in the proof of Lemma (11.11) If we put $\tilde{F}=F|(\tilde{V}\setminus\{0\})$ and choose a point $\tilde{x}\in p^{-1}(x)$, we get a commutative diagram

$$\begin{array}{ccc} (\tilde{V},\tilde{x}) & \xrightarrow{\tilde{F}_{\tilde{x}}} & (\mathbb{C}^{n+1}\setminus\{0\},\ \tilde{F}(\tilde{x})) \\ p\downarrow & & \downarrow p \\ (V,x) & \xrightarrow[\varphi_x]{} & (\mathbb{P}^n,\ \varphi(x)) \end{array}$$

A moment thought shows that the germ $\tilde{F}_{\tilde{x}}$ is a trivial unfolding of the germ φ_x, in other words $\tilde{F}_{\tilde{x}}=\varphi_x\times \mathrm{id}$, where id denotes the germ $(\mathbb{C}^1,0)\to(\mathbb{C}^1,0)$ of the identity map.

It follows easily from this remark the equality of Boarmand symbols $\Sigma\tilde{F}_{\tilde{x}}=\Sigma\varphi_x$. Moreover, since $\tilde{V}$ is a smooth hypersurface at the point $\tilde{x}$, it follows that $\Sigma\tilde{F}_{\tilde{x}}=\Sigma I$, where I is the ideal in ${}_{\tilde{x}}E_{n+1}$ generated by the germs f and $\frac{\partial f}{\partial x_i}-\frac{\partial f}{\partial x_i}(\tilde{x})$ for $i=0,\dots,n$. By (11.29.ii) and (11.30) it follows that $K_x(V)=A_2$ if and only if $\Sigma I=(1,0,\dots)$.

Note also the explicit description of the first critical Jacobian extension

$$\Delta^1 I = I + (H(f)) \subset {}_{\tilde{x}}E_{n+1} .$$

By a direct computation, it follows that $\Sigma I=(1,0,0,0,\ldots)$ if and only if the matrix having as rows the differentials $dH(f)(\tilde{x})$ and $d(\frac{\partial f}{\partial x_i})(\tilde{x})$ for $i=0,\ldots,n$ has rank $n+1$.

Recalling that $x \in \mathrm{Inf}(V)$ and using again the Euler relation for the derivatives $\frac{\partial f}{\partial x_i}$ as in the proof of (11.11) we get the result.

(11.35) EXERCISE

Show that a point $x \in V$ with corank $K_x(V) \geq 2$ is necessarily a singular point for the hessian hypersurface $H(V)$. Hint: use the rule for differentiating a determinant.

§3. HYPERPLANE SECTIONS OF GENERIC HYPERSURFACES

A classical result in Algebraic Geometry is *Bertini's Theorem* which says that for a smooth variety $X \subset \mathbb{P}^n$ and for a generic hyperplane $H \subset \mathbb{P}^n$, the hyperplane section $X \cap H$ is again a smooth variety (e.g. [Ha], p. 179).

In this section we want to discuss a phenomenon with a similar flavour: if we start with a generic smooth hypersurface V in $\mathbb{P}^n$ and consider all of its hyperplane sections $V \cap H$, then something can be said about the singularities these sections can acquire.

For instance let K be a contact equivalence class in m_{n-1}^2 and, for a smooth hypersurface V in $\mathbb{P}^n$, put

$$K(V)=\{x \in V;\quad K_x(V)=K\} .$$

Then, fixing n and $d=\deg(V)$, one may ask which are the contact equivalence classes K such that $K(V) \neq \emptyset$ for a generic hypersurface V.

To answer such questions, we need the following framework.

Let $P_o(n,d)$ be the vector space of homogeneous polynomials f in $x_o,\dots,x_n$ of degree d and $U\subset P_o(n,d)$ be the Zariski open subset of polynomials f such that $V(f):f=0$ is a smooth hypersurface in $\mathbb{P}^n$ (compare to (7.15)).

We have a linear surjective map

$$r:P_o(n,d)\to P_o(n-1,d),\quad r(f)=f\big|_{x_n=0}$$

and a linear isomorphism

$$h:P_o(n-1,d)\to \mathbb{C}\times J^d(n-1,1);\quad h(f)=(f(a),\ f(1,u_1,\dots,u_{n-1})-f(a))$$

where $a=(1,0,\dots,0)$ and $u_1,\dots,u_{n-1}$ denote the coordinates on $\mathbb{C}^{n-1}$.

Let $q=h\circ r$ be the composition of these two linear maps. Note that for $f\in U$ and for a contact equivalence class $K\subset m^2_{n-1}$, one has the next equivalent statements.

(11.36) (i) $a\in V(f)$, $T_aV(f):x_n=0$ and $K_a(V(f))=K$.

(ii) $q(f)=(0,g)$ and $g=0$ defines an isolated singularity at the origin of $\mathbb{C}^{n-1}$ of type K or, in other words, $g\in K$.

For a contact class $K\subset m^2_{n-1}$ corresponding to a function germ g as above, we introduce the next notations:

(a) $\mu(K)=\mu(g)$, $\tau(K)=\tau(g)$ - the Milnor and Tjurina numbers;

(b) $O(K)=\min\{s;\ m^s_{n-1}\subset T\mathcal{K}f\}$ - the order of strong $\mathcal{K}$-determinacy by Theorem (6.27.i) .

(c) $S^d_K=\{\tilde{g}\in J^d(n-1,1);\ \tilde{g}\in K\}$

where $\tilde{g}$ is regarded as a polynomial function germ. When $d\geq O(K)$ it is clear that $S^d_K=\mathcal{K}^d_{n-1,1}\cdot j^dg$ for some $g\in K$ and hence in this case S^d_K is a smooth semialgebraic irreducible variety in $J^d(n-1,1)$ of codimension equal to

$$\operatorname{codim} S^d_K=\tau(K)+n-2$$

by (6.52.ii) .

(d) If K' is a different contact equivalence class in m^2_{n-1}, then we say that K is d-*adjacent* to K' when $S^d_K\subset\overline{S^d_{K'}}$.

Now we can start the basic construction for describing the generic contact tangency classes. Let us consider the incidence relation

$$M=\{(x,H)\in \mathbb{P}^n\times\hat{\mathbb{P}}^n;\ x\in H\}$$

It is easy to see that M is a smooth projective variety with dim M=2n-1.

On the other hand, note that $\mathbb{C}^*$ acts by multiplication on the vector space $P_o(n,d)$ in such a way that the open set U is $\mathbb{C}^*$-invariant. Let P(n,d) be the corresponding quotient space $U/\mathbb{C}^*$, in fact an open subset in the projective space $\mathbb{P}(P_o(n,d))$.

For a contact class $K\subset m^2_{n-1}$ we define

$$(11.37)\qquad Z_K=\{(f,x,H)\in P(n,d)\times M;\ x\in K(V(f))\ \text{and}\ T_xV(f)=H\}$$

(11.38) LEMMA

Z_K is a semialgebraic subset in P(n,d)×M of dimension

$$\dim Z_K=\dim P(n,d)+2n-2-\operatorname{codim} S^d_K\ .$$

Moreover, for d≥O(K), Z_K is smooth and irreducible.

PROOF

Note that the map $p_2:Z_K\to M$ induced by the second projection is a locally trivial fiber bundle with fiber

$$F=q^{-1}(0\times S^d_K)/\mathbb{C}^*\ .$$

All the statements follow easily from this remark.

Next note that $K(V(f))=p_1^{-1}(f)\cap Z_K$, where $p_1:P(n,d)\times M\to P(n,d)$ is the first projection. It follows that K(V(f)) is a *semialgebraic subset* in the hypersurface V(f) and it is *nonempty* if and only if $f\in p_1(Z_K)$.

Moreover, note that the intersection $p^{-1}(f)\cap Z_K$ is *transverse* at a point $(f,m)\in P(n,d)\times M$ if Z_K is smooth and $p_1:Z_K\to P(n,d)$ is a submersion at the point (f,m).

These simple remarks justify the next

(11.39) DEFINITION

A contact tangency class $K \subset m_{n-1}^2$ is called *generic for the pair* (n,d) if there is a dense Zariski open subset $W \subset P(n,d)$ such that $f \in W$ implies $K(V(f)) \neq \emptyset$.

(11.40) LEMMA

The following statements are equivalent.

(i) K *is generic for the pair* (n,d).

(ii) $p_1(\overline{Z}_K) = P(n,d)$.

(iii) $\overline{p_1(Z_K)} = P(n,d)$, *in other words* $p_1 : Z_K \to P(n,d)$ *is a dominant morphism.*

(iv) $\dim p_1(Z_K) = \dim P(n,d)$.

PROOF

Simple exercise for the reader using the irreducibility of P(n,d).

Note that $\overline{Z}_K$ is a fiber bundle over M with fiber $\overline{F} = \bigcup q^{-1}(0 \times S_{K'}^d)/\mathbb{C}^*$, the union being over all the contact classes K' which are d-adjacent to K.

Hence (11.40.ii) says that any smooth hypersurface V(f) has at some point x a contact tangency class which is d-adjacent to K.

The main result of this section is the following.

(11.41) THEOREM

Let K *be a contact tangency class in* m_{n-1}^2 *and let* $d \geq O(K)$ *be a positive integer. Then* K *is generic for the pair* (n,d) *if and only if* $n - \tau(K) \geq 0$.

Moreover, if K *is generic for the pair* (n,d), *then there is a Zariski open and dense subset* $W \subset P(n,d)$ *such that* $f \in W$ *implies that* K(V(f)) *is a smooth nonempty algebraic variety of dimension*

$$\dim K(V(f)) = n - \tau(K)$$

PROOF

If K is generic for the pair (n,d) we clearly have $\dim Z_K \geq \dim P(n,d)$ and using (11.38) this gives $n-\tau(K)\geq 0$.

Conversely, using the theorem on the dimension of the fibers of a morphism (e.g. [Ha], p. 95), it is enough to find a polynomial $f \in P(n,d)$ such that for some point $x \in K(V(f))$ one has

$$\dim_x K(V(f)) = n-\tau(K) \ .$$

Here, by $\dim_x A$ we mean the dimension of the analytic set germ (A,x).

This part of the proof depends heavily on Bruce's approach in [Bc], p. 39 where the corresponding local smooth situation is treated.

STEP 1: CONSTRUCTION OF THE EQUATION

Take $g \in J^d(n-1,1)$ be a sufficient jet in the contact class K. Then $TKg \supset m^d$, where m will denote in this proof the maximal ideal m_{n-1}.

Let r=corank (g) and use Splitting Lemma to write

$$g(x,y) = \bar{g}(x) + y_{r+1}^2 + \ldots + y_{n-1}^2$$

where $x=(x_1,\ldots,x_r)$, $y=(y_{r+1},\ldots,y_{n-1})$ and $j^2\bar{g}=0$. Then $\tau(g) = \tau(\bar{g})$ and let τ denote this common Tjurina number.

Let $\{1,x_1,\ldots,x_r,u_{r+1},\ldots,u_{\tau-1}\}$ be a monomial basis for the Tjurina algebra $T(\bar{g})$. We clearly have

$$1 < \deg u_j(x) < d \quad \text{for} \quad j=r+1,\ldots,\tau-1 \ .$$

Consider the following modification of the function germ g:

$$\tilde{g}(x,y) = \bar{g}(x) + \sum_j (y_j + u_j(x))^2 + \sum_k y_k^2$$

where $j=r+1,\ldots,\tau-1$, $k=\tau,\ldots,n-1$ and if $\tau=n$ this last sum is omitted.

Let $h=j^d\tilde{g}$, a jet obtained from $\tilde{g}$ by deleting all the squares $u_j^2(x)$ of degree $>d$.

The coordinate change $\bar{y}_j = y_j + u_j(x)$ shows that $\tilde{g}$ is contact equivalent to g and hence $TK\tilde{g} \supset m^d$. This implies that h is contact equivalent to $\tilde{g}$ and that $TKh = TK\tilde{g}$.

Moreover, by Theorem (6.27.i)) it follows that h is strongly d-K-determined. Hence we can and do modify slightly the terms of degree d in h such that to ensure that the equation $h_d=0$ defines a smooth hypersurface in $\mathbb{P}^{n-2}$, where h_d denotes the sum of these highest degree terms in h.

Consider now an affine hypersurface X in $\mathbb{C}^n$ given by an equation of the form

(11.42) $\quad X: z+tz^d=h(x,y) \qquad t \in \mathbb{C}^*.$

STEP 2: SMOOTHNESS OF THE HYPERSURFACE X

We claim that for a generic choice of the parameter t, the hypersurface X has a smooth projective closure $\bar{X}$ in $\mathbb{P}^n$.

If $t \neq 0$, then the intersection of $\bar{X}$ with the hyperplane at the infinity $H = \mathbb{P}^n \setminus \mathbb{C}^n \simeq \mathbb{P}^{n-1}$ is given by the equation

$$tz^d = h_d(x,y)$$

and this is a smooth hypersurface in H by our remark above on h_d.

On the other hand, a simple argument shows that the function $h: \mathbb{C}^{n-1} \to \mathbb{C}$ has a finite number of singular points. (Hint: use again the property of h_d.) Hence h has a finite number of critical values.

Next, the critical values of the function $z+tz^d$ depend continuously on t and hence for a generic value of t these values are different from the critical values of h.

For such a choice, X is clearly smooth. The two smoothness facts proved above show that $\bar{X}$ is a smooth hypersurface in $\mathbb{P}^n$ of degree d, for a generic value of the parameter t. In the sequel we will take also t sufficiently small.

Let $\bar{x} \in \bar{X}$ be the point corresponding to the origin 0 of $\mathbb{C}^n$.

STEP 3: $\bar{x} \in K(\bar{X})$ *and* $\dim_{\bar{x}} K(\bar{X}) = n-\tau$.
Since the notion of contact tangency class can be defined (similarly and compatibly!) for affine hypersurfaces, it is enough to work with the hypersurface X.

Note that $T_oX:z=0$ and hence, identifying T_oX with the (x,y)-hyperplane, the singularity $(T_oX \cap X,0)$ is given by $h(x,y)=0$. This implies that $0 \in K(X)$, since $h \in K$ by its construction, and this proves the first part of our claim.

Note that the hyperplanes in a neighbourhood of T_oX can be parametrized by $a \in \mathbb{C}^{n-1}$ (a small enough) by the explicit equation

$$H_a: \sum_i a_i x_i + \sum_j a_j y_j + z = 0$$

To investigate the position of these hyperplanes with respect to the hypersurface X (in the neighbourhood of the origin) it is natural to consider the map germ

$$(11.43) \quad H:(\mathbb{C}^{n-1} \times \mathbb{C}^{n-1},(0,0)) \to (\mathbb{C},0)$$

$$H(x,y,a) = \sum_i a_i x_i + \sum_j a_j y_j + \bar{h}(x,y)$$

where the function germ $\bar{h}$ is obtained as follows.

Since $T_oX:z=0$, it follows that the hypersurface germ $(X,0)$ can be given by an equation $z=\bar{h}(x,y)$. Moreover, note that $\bar{h}(x,y)-h(x,y) \in m^d$ and the terms of degree d in this difference have coefficients arbitrarly small, if we take t small enough.

By the strong d-K-determinacy of h, if follows that $\bar{h}$ is contact equivalent to h and that $TK\bar{h}=TKh$.

A word of explanation about the map germ H is now in order. If we fix a and look at the function germ $H(\ ,a):(\mathbb{C}^{n-1},0) \to (\mathbb{C},0)$ then a point $(x,y) \in \mathbb{C}^{n-1}$ is a singularity of contact equivalence class K' for this function if and only if the hyperplane H_a is the tangent hyperplane to the hypersurface X at the point $p=(x,y,\bar{h}(x,y))$ and $p \in K'(X)$.

To investigate the germ $(K(X),0)$ using this remark, we consider the local jet map

$$(11.44) \quad j_1^d H:(\mathbb{C}^{n-1} \times \mathbb{C}^{n-1},(0,0)) \to (J^d(n-1,1),\ j_1^d H(0,0))$$

where $j_1^d H(x,y,a)$ is the d-jet of the function

$$(u,v) \to H(x+u,y+v,a)-H(x,y,a)$$

In particular $j_1^d H(0,0)=j^d\bar{h}$.

Note that, modulo m^d, we have the equality

$$H(x,y,a)=\sum_i a_i x_i+\sum_j a_j y_j+\tilde{g}(x,y)$$

Then, modulo m^d, the image of the differential at the origin of the map $j_1^d H$ is spanned by the vectors

$$\{x_1,\dots,x_r;y_{r+1},\dots,y_{n-1};\frac{\partial\tilde{g}}{\partial x_i} \text{ for } i=1,\dots,r;\ u_j(x) \text{ for } j=r+1,\dots,\tau-1\}$$

as in [Bc], p. 39.

It is easy to see that $\{1,x_1,\dots,x_r,u_{r+1},\dots,u_{\tau-1}\}$ give a basis for the Tjurina algebra $T(\tilde{g})\simeq T(\bar{g})$. This implies the last equality below:

$$\langle x_i\rangle+\langle y_j\rangle+\langle\frac{\partial\tilde{g}}{\partial x_i}\rangle+\langle u_j\rangle+TK\tilde{g}=\langle x_i\rangle+\langle u_j\rangle+\langle\frac{\partial\tilde{g}}{\partial x_i}\rangle+\langle\frac{\partial\tilde{g}}{\partial y_j}\rangle+TK\tilde{g}=m$$

Recalling now the equalities $TK\tilde{g}=TKh=TK\bar{h}$, it follows that $j_1^d H$ is transverse to the (sufficient) contact orbit Y of $j^d\bar{h}$ in $J^d(n-1,1)$.

Now codim Y in $J^d(n-1,1)$ is precisely $n+\tau-2$ by (6.52). On the other hand $((j_1^d H)^{-1}(Y),0)$ identifies to the germ $(K(X),0)$ by the discussion above. It follows that

$$\dim\ (K(X),0)=\dim\ ((j_1^d H)^{-1}(Y),0)=2n-2-(n+\tau-2)=n-\tau$$

and this ends the main part of the proof of Theorem (11.41).

To prove the last statement in (11.41), we note that for a generic contact tangency class K we have a dominant morphism $p_1:Z_K\to P(n,d)$. It follows from Sard's Theorem [Mu], p. 42 that there is a Zariski open and dense subset $W\subset P(n,d)$ such that p_1 is a submersion at any point $z\in Z_K$ with $f=p_1(z)\in W$.

But then $p_1^{-1}(f)\cap Z_K$ is a transverse intersection at z and this ends the proof.

(11.45) COROLLARY

(i) *For plane curves of degree* $d\geq 3$, *the generic contact tangency classes are* A_1 *and* A_2.

(ii) *For surfaces in* $\mathbb{P}^3$ *of degree* $d\geq 3$, *the generic contact tangency classes are* A_1 , A_2 *and* A_3.

PROOF

(i) Since $n=2$ and $d\geq 3$, everything follows from (11.41).

(ii) Now $n=3$ and for $d\geq 4=O(A_3)$ the result follows from (11.41). The special case $d=3$ is left as an exercise for the reader.

(11.46) EXERCISE

Use (11.45.i) and (11.34) to give a new proof for Corollary (11.20). Explain why any smooth cubic curve has exactly nine inflexion points.

Finally we want to consider a little bit *the set of all singular points of a hyperplane section* $V_o=V\cap H$ of a smooth projective hypersurface V. We start with the following.

(11.47) EXAMPLE

Let $H=\mathbb{P}^2\subset\mathbb{P}^3$ and consider d lines in H, no three of them meeting at a point. Then their union forms a singular curve V_o of degree d having exactly $\binom{d}{2}=\frac{d(d-1)}{2}$ nodes, the intersection points of the lines.

By Proposition (11.6), there is a smooth surface V in $\mathbb{P}^3$ of degree d such that $V_o=V\cap H$.

This example shows that a hyperplane can be tangent to a smooth hypersurface at a lot of points.

On the other hand, for generic hypersurfaces this phenomenon does not occur. Before giving a precise statement, we need the next

(11.48) DEFINITION

The points $a_1,\dots,a_k\in\mathbb{P}^n$ are said to be *in general position* if the vector subspace in $\mathbb{C}^{n+1}$ generated by the corresponding lines $\mathbb{C}a_1,\dots,\mathbb{C}a_k$ has dimension k.

In particular, note that $k\leq n+1$.

(11.49) PROPOSITION

For a generic hypersurface $V \subset \mathbb{P}^n$, *the singular points of any hyperplane section* $V_o = V \cap H$ *are in general position in* $H \simeq \mathbb{P}^{n-1}$.

In particular, a plane $H \subset \mathbb{P}^3$ can be tangent to a generic surface $S \subset \mathbb{P}^3$ at maximum 3 (non-collinear!) points.

A proof of Proposition (11.49) can be found in [Bc] or [D2].

There is also an analogue of Theorem (11.41) involving the set of all singularities in the hyperplane sections and such type of results are necessary for understanding the structure of the non-isolated singularities of the dual hypersurfaces. For more details, we refer the interested reader to Bruce's paper [Bc].

APPENDIX

TWO RECENT RESEARCH TOPICS

We point out here (mostly by indicating the appropriate references) two interesting recent directions of research which are closely related to the topics discussed in this book. Moreover, it is quite likely that these directions will yield some major new results in a not too distant future.

The first of these topics consists in investigating the simplest *non-isolated* singularities of function germs $f:(\mathbb{C}^n,0) \to (\mathbb{C},0)$, namely those germs f for which the associated hypersurface singularity $X:f=0$ has a 1-dimensional singular space (i.e. dim Sing $(X)=1$).

This direction was initiated by D. Siersma and some useful references here are the following.

1. Siersma, D.: Isolated line singularities, PROC. SYMP. PURE MATH. 40 (ARCATA, PART II), Amer. Math. Soc. (1983), 485--496.
2. Siersma, D.: Quasihomogeneous singularities with transversal type A_1 , preprint Utrecht University, 1987.
3. Pellikaan, R.: Hypersurface singularities and resolutions of Jacobi modules, Thesis Utrecht University, 1985.

In addition, the very recent preprint [2] contains a more extensive reference list for this topic. The second recent research direction consists in passing from the study of function germs $f:(\mathbb{C}^n,0) \to (\mathbb{C},0)$ to the more general study of function germs $f:(X,0) \to (\mathbb{C},0)$ *defined on an analytic space germ* $(X,0)$. There are many motivations for this extension. One of them can be found in [D3], some others in the next references.

The simplest case to handle is when $(X,0)$ is an isolated hypersurface (or, more generally, complete intersection singularity) and it was V.I. Arnold who has originally insisted upon the necessity of such a study. As references, we mention

4. Lyashko, O.V.: Classification of critical points of functions on a manifold with singular boundary, Funct. Anal. Appl. 17 (1983), 187-193.

5. Dimca, A.: Function germs defined on isolated hypersurface singularities, Compositio Math. 53 (1984), 245-258. See also [D3].

The treatment in [5] is very close to that in this book and the reader will encounter no difficulties to understand all the details. Moreover, he will find out there that the $\mathcal{R}$-simple function germs do no longer coincide to the $\mathcal{K}$-simple ones in this more general setting (and that even the existence of simple function germs is a problem!).

Another interesting case to consider is when (X,0) is a quotient singularity and this is done in

6. Wall, C.T.C.: Functions on quotient singularities, preprint Liverpool University, 1986.

The general case when (X,0) is a complex analytic set germ was recently treated with more sophisticated tools in

7. Bruce, J.W. and Roberts, R.M.: Critical points of functions on analytic varieties, preprint Warwick University, 1986.

This last preprint contains also a more complete reference list on the subject.

There are many possibilities now for starting a research work in singularities. We hope that our suggestions above may help a little a beginner who misses a competent advisor.

REFERENCES

[Ad] Adams, J.F.: LECTURES ON LIE GROUPS, W,A, Benjamin Inc., New York-Amsterdam, 1969.

[AGV] Arnold, V.I., Gusein-Zade, S.M. and Varchenko, A.N.: SINGULARITIES OF DIFFERENTIABLE MAPS I, Birkhauser, Basel-Boston, 1985.

[Ar] Artin, M.: On the solutions of analytic equations, Invent. Math. 5 (1968), 277-291.

[AM] Atiyah, M.F. and Macdonald, I.G.: INTRODUCTION TO COMMUTATIVE ALGEBRA, Addison-Wesley, Reading-Massachusetts, 1969.

[BPV] Barth, W., Peters, C. and Van de Ven, A.: COMPACT COMPLEX SURFACES, Springer, Berlin 1984.

[BS] Briançon, J. and Skoda, H.: Sur la clôture intégrale d'un idéale de germes de fonctions holomorphes en un point de $\mathbb{C}^n$, C.R. Acad. Sci. Paris 278 (1974), 949-951.

[B] Brieskorn, E.: Singular elements of semisimple algebraic groups, in Actes Congr. Int. Math. Nice 2 (1970), 279--284.

[BK] Brieskorn, E. and Knorrer, H.: PLANE ALGEBRAIC CURVES, Birkhauser, Basel-Boston, 1986.

[Bc] Bruce, J.W.: The duals of generic hypersurfaces, Math. Scand. 49 (1981), 36-60.

[BG] Bruce, J.W. and Giblin, P.J.: CURVES AND SINGULARITIES, Cambridge Univ. Press, Cambridge, 1984.

[CD] Choudary, A.D.R. and Dimca, A.: On the dual and hessian mappings of projective hypersurfaces, Math. Proc. Camb. Phil. Soc. 101 (1987), 461-468.

[Da] Damon, J.: The classification of discrete algebra types ($n \leq p$), unpublished preprint.

[D1] Dimca, A.: A geometric approach to the classification of pencils of quadrics, Geom. Dedicata 14 (1983), 105--111.

[D2] Dimca, A.: Tangencies of generic real projective hypersurfaces, Math. Scand. 53 (1983), 216-220.

[D3] Dimca, A.: Are the isolated singularities of complete intersections determined by their singular subspaces?, Math. Ann. 267 (1984), 461-472.

[D4] Dimca, A.: Milnor numbers and multiplicities of dual varieties, Rev. Roumaine Math. Pures Appl. 31 (1986), 535-538.

[DG1] Dimca, A. and Gibson, C.G.: On contact simple germs of the plane, preprint INCREST no. 20 (1981).

[DG2] Dimca, A. and Gibson, C.G.: On contact germs of the plane, PROC. SYMP. PURE MATH. 40 (ARCATA, PART I), Amer. Math. Soc. (1983), 277-282.

[DG3] Dimca, A. and Gibson, C.G.: Contact unimodular germs from the plane to the plane, Quart. J. Math. Oxford 34 (1983), 281-295.

[DG4] Dimca, A. and Gibson, C.G.: Classification of equidimensional contact unimodular map germs, Math. Scand. 56 (1985), 15-28.

[Do] Dolgachev, I.: Weighted projective varieties, in GROUP ACTIONS AND VECTOR FIELDS, Lecture Notes in Math. 956, Springer, Berlin, 1982, 34-71.

[Du] Durfee, A.H.: Fifteen characterizations of rational double points and simple critical points, L'Enseignement Math. 25 (1979), 131-163.

[F] Fischer, G.: COMPLEX ANALYTIC GEOMETRY, Lecture Notes in Math. 538, Springer, Berlin, 1976.

[GH] Gaffney, T. and Hauser, H.: Characterizing singularities of varieties and of mappings, Invent. Math. 81 (1985), 427-447.

[Gi] Gibson, C.G.: SINGULAR POINTS OF SMOOTH MAPPINGS, Research Notes in Math. 25, Pitman, London, 1979.

[GWPL] Gibson, C.G.; Wirthmüller, K.; du Plessis, A.A. and Looijenga, E.J.N.:TOPOLOGICAL STABILITY OF SMOOTH MAPPINGS, Lecture Notes in Math. 552, Springer, Berlin, 1977.

[Gt] Giusti, M.: Classification des singularités isolées d'intersections complètes simples, C.R. Acad. Sci. Paris, 284 (1977), 167-170; see also PROC. SYMP. PURE MATH. 40 (ARCATA, PART I), Amer. Math. Soc. (1983), 457-494.

[GG] Golubitsky, M. and Guillemin, V.: STABLE MAPPINGS AND THEIR SINGULARITIES, Graduate Texts in Math. 14, Springer, Berlin-New York, 1973.

[Ha] Hartshorne, R.: ALGEBRAIC GEOMETRY, Graduate Texts in Math. 52, Springer, Berlin-New York, (third printing), 1983.

[HP] Hodge, W. and Pedoe, D.: METHODS OF ALGEBRAIC GEOMETRY, Cambridge Univ. Press, Cambridge, 1947.

[H1] Humphreys, J.E.: INTRODUCTION TO LIE ALGEBRAS AND REPRESENTATION THEORY, Graduate Texts in Math. 9, Springer, Berlin-New York, 1972.

[H2] Humphreys, J.E.: LINEAR ALGEBRAIC GROUPS, Graduate Texts in Math. 21, Springer, Berlin-New York, 1975.

[I] Iarrobino, A.A.: PUNCTUAL HILBERT SCHEMES, Memoirs Amer. Math. Soc. 188 (1977).

[Is] Ishii, S.: A Characterization of hyperplane cuts of smooth complete intersections, Proc. Japan Acad. 58 (1982), 309-311.

[KK] Kaup, B. and Kaup, L.: HOLOMORPHIC FUNCTIONS OF SEVERAL VARIABLES, de Gruyter Studies in Math. 3, Berlin-New York, 1983.

[Kl] Kleiman, S.L.: The enumerative theory of singularities, in REAL AND COMPLEX SINGULARITIES (OSLO 1976), Sijthoff and Noordhoff, 1977, 297-396.

[Kn] Knörrer, H.: Isolierte singularitäten von Durchschnitten zweier Quadriken, Bonner Math. Schriften 117 (1980).

[Lm] Lamotke, K.: REGULAR SOLIDS AND ISOLATED SINGULARITIES, Vieweg Advanced Lectures in Math., Vieweg, Braunschweig, 1986.

[L] Lang, S.: ALGEBRA, Addison-Wesley, Reading-Massachustts, 1965.

[La] Laufer, H.: NORMAL TWO-DIMENSIONAL SINGULARITIES, Annals of Math. Studies 71, Princeton Univ. Press, Princeton, 1971.

[Ło] Łojasiewicz, S.: ENSEMBLES SEMI-ANALITIQUES, IHES Lecture Notes, 1965.

[Ln] Looijenga, E.J.N.: ISOLATED SINGULAR POINTS ON COMPLETE INTERSECTIONS, London Math. Soc. Lecture Note Series 77, Cambridge Univ. Press, Cambridge, 1984.

[Mr] Martinet, J.: SINGULARITIES OF SMOOTH FUNCTIONS AND MAPS, London Math. Soc. Lecture Note Series 58, Cambridge Univ. Press, Cambridge, 1982.

[Ma] Mather, J.N.: Stability of C^∞-mappings IV: Classification of stable germs by $\mathbb{R}$-algebras, Publ. Math. IHES 37 (1970), 223-248.

[MY] Mather, J.N. and Yau, S.S.T.: Classification of isolated hypersurface singularities by their moduli algebras, Invent. Math. 69 (1982), 243-251.

[M1] Milnor, J.: SINGULAR POINTS OF COMPLEX HYPERSURFACES, Annals of Math. Studies 61, Princeton Univ. Press, Princeton, 1968.

[M2] Milnor, J.: MORSE THEORY, Annals of Math. Studies 76, Princeton Univ. Press, Princeton, 1974.

[Mo] Montaldi, J.A.: On contact between submanifolds, Michigan Math. J. 33 (1986), 195-199.

[Mu] Mumford, D.: ALGEBRAIC GEOMETRY I: COMPLEX PROJECTIVE VARIETIES, Springer, Berlin-New York, 1976.

[Ph] Pham, F.: Courbes discriminantes des singularités planes d'ordre 3, in SINGULARITÊS À CARGÊSE 1972, Astérisque 7-8, Soc. Math. France (1973), 363-391.

[P] Pinkham, H.: Singularités rationnelles de surfaces-Appendice, in SEMINAIRE SUR LES SINGULARITES DES SURFACES, Lecture Notes in Math. 777, Springer, Berlin, 1980.

[R] Reichard, K.: Nichtdifferenzierbare Morphismen differenzierbarer Räume, Manuscripta Math. 15 (1975), 243--250.

[S1] Saito, K.: Quasihomogene isolierte Singularitäten von Hyperflächen, Invent. Math. 14 (1971), 123-142.

[S2] Saito, K.: Einfach-elliptische Singularitäten, Invent. Math. 23 (1974), 289-325.

[Si] Siersma, D.: Periodicities in Arnold's lists of singularities, in REAL AND COMPLEX SINGULARITIES (OSLO, 1976), Sijthoff and Noordhoff, 1977, 497-524.

[Te] Teissier, B.: Cycles évanescents, sections planes et conditions de Whitney, in SINGULARITES A CARGESE 1972, Asterisque 7-8, Soc. Math. France (1973), 285-362.

[T] Tougeron, J.C.: IDEAUX DE FONCTIONS DIFFERENTIABLES, Ergebnisse der Math. 71, Springer, New York, 1972.

[W1] Wall, C.T.C.: A second note on symmetry of singularities, Bull. London Math. Soc. 12 (1980), 347-354.

[W2] Wall, C.T.C.: Finite determinacy of smooth map germs, Bull. London Math. Soc. 13 (1981), 481-539.

[W3] Wall, C.T.C.: Classification of unimodal isolated singularities of complete intersections, PROC. SYMP. PURE MATH. 40 (ARCATA, PART II), Amer. Math. Soc. (1983), 625-640.

[Ws] Wassermann, G.: STABILITY OF UNFOLDINGS, Lecture Notes in Math. 393, Springer, Berlin, 1974.

[Wt] Whitney, H.: COMPLEX ANALYTIC VARIETIES, Addison-Wesley, Reading-Massachusetts, 1972.

The reader can find additional useful references in [AGV], [Du], [Ln] and [W2], the first one giving a good idea about the substantial contributions of the Russian School in singularities.

LIST OF NOTATIONS

INDEX